MATHEMATICS RESEARCH DEVELOPMENTS

PSEUDO-MATROIDS AND CUTS OF MATROIDS

MATHEMATICS RESEARCH DEVELOPMENTS

Additional books in this series can be found on Nova's website under the Series tab.

Additional e-books in this series can be found on Nova's website under the eBook tab.

MATHEMATICS RESEARCH DEVELOPMENTS

PSEUDO-MATROIDS AND CUTS OF MATROIDS

SERGEY A. GIZUNOV
AND
V. N. LYAMIN

nova publishers
New York

Library of Congress Cataloging-in-Publication Data

Names: Gizunov, Sergey A., editor. | Lyamin, V. N., editor.
Title: Pseudo-matroids and cuts of matroids / Sergey A. Gizunov and V.N. Lyamin (Scientific Research Institute "KVANT", Moscow, Russia), editors.
Description: Hauppauge, New York : Nova Science Publishers, Inc., [2016] | Series: Mathematics research developments | Includes index.
Identifiers: LCCN 2016007433 (print) | LCCN 2016013622 (ebook) | ISBN 9781634848817 (hardcover) | ISBN 9781634848978 ()
Subjects: LCSH: Matroids. | Combinatorial designs and configurations. | Graph theory. | Linear dependence (Mathematics) | Algebras, Linear.
Classification: LCC QA166.6 .P74 2016 (print) | LCC QA166.6 (ebook) | DDC 511/.6--dc23
LC record available at http://lccn.loc.gov/2016007433

Published by Nova Science Publishers, Inc. † New York

CONTENTS

FOREWORD

The matroid theory was begun in the 1930s, when B.L. Van der Waerden in his work "Modern Algebra" examined algebraic dependance, along with linear dependance, and H. Whitney (H. Whitney, 1935) introduced the abstract notion of a matroid, in order to generalize the notion of the dual graph. Later on, S. MacLane (S. MacLane, 1936) gave an interpretation of matroids in terms of projective geometry (which served as a reason for matroids to be referred to as combinatory geometrics), while G. Birkhoff (G. Birkhoff, 1935) introduced the notion of the "*M*-structure" and noted that it includes projective geometries.

Matroids appear in various combinatory algebraic contexts. Such notions as independence and basis in vector spaces, algebraic dependance, cycles and cuts in graphs, surfaces in projective geometries, and point semi-modular lattices all come down to the structure of the matroid. Owing to the possibility of implementing the lattice theory, graphs, vector spaces and geometry language in the description of matroids, some unexpected similarities between the results of graph theory, coding theory, algebra, topology, electro-technics, geometry and combinatorial mathematics were found.

Matroids *M* and *N*, defined on the set *S*, are related by a strong mapping, if there is a one-to-one mapping $\tau : S \to S$, such that the pre-image of every closed set of the matroid *N* is a closed set of the matroid *M*, and are related by a weak mapping, if the pre-image of every independent set of the matroid N is an independent set of the matroid M under one-to-one mapping τ.

This study examines both strong and weak mappings of matroids. The notion of the "pseudo-matroid" is introduced as a particular type of matroid generated by elementary strong mappings. Various properties of pseudo-matroids are studied. It is remarkable that pseudo-matroids unequivocally define the mappings by which they are generated. This property has allowed the authors of the study to obtain an array of fresh results.

For a family of binary matroids the notion of "G-mappings" is introduced. Characteristics of pseudo-matroids, generated by G-mappings (which the authors refer to as semi-matroids), are examined. It is proven that the class of the binary matroid is closed under G-mappings and rank-preserving weak mappings.

An algorithm for building the families of cycles of all non-isomorphic binary matroids on finite set is offered. All non-isomorphic binary matroids on the set S, where $1 \leq |S| \leq 15$, are enumerated. Also, the results obtained in this study make it possible to enumerate all non-isomorphic linear codes over the set S, where $1 \leq |S| \leq 22$.

Pseudo-matroids, introduced by the authors of this work, and their properties present an instrument for further studies of matroid categories and mappings.

Professor A. M. Revyakin, PhD

PREFACE

The term "matroid" was coined by H. Whitney [15] in the 1930s in a work dedicated to the development of Van der Waerden's idea of the possibility of a general theoretical approach to the study of both algebraic and linear dependences. As the theory developed, the scope of matroid dependences broadened with the inclusion of other combinatorial objects, such as graphs, transversal curves and block designs. This fact has not only confirmed the backbone character of matroid dependences, but has also allowed the joining of a scope of directions and classes of combinatorial analysis conceptually into one general theory, though these previously seemed diverse and non-related. A substantial leap in the development of this theory was its categorization, i.e., the introduction of morphisms between matroids, more specifically of strong and weak mappings. The existence of strong and weak mappings allows for the introduction of partial order over the set of all matroids, where overlaps would be uniquely defined by some algebraic structures.

The author's research [16-18] has shown that the elements of such structures, minimal by inclusion, are, in turn, part of more general algebraic objects of matroid type, defined in this work as pseudo-matroids, generated by elementary (strong) mappings of matroids. This work is thematic and is mainly dedicated to the study of the properties of such objects.

It is worth noting that within more than 70 years of history of the general algebraic theory of matroids the appropriateness of its establishment as a separate branch of discrete mathematics, which would be dedicated to the analysis of algebraic dependences of a certain type, has been proven.

At the moment, the fullest and most systematic presentation of matroid theory, the authors believe, is offered in the monographs of Welsh [14] and Oxley [12], published in 1976 and 1992, respectively. Oxley's monograph with additions was republished in 2006. "Introduction to the theory of matroids" by Tutte [13] is also worth mentioning, as well as a 1970 article on matroids by Crapo and Rota [3] from a series under the title of "On the Foundations of Combinatorial Theory". In this article matroids are studied primarily as ordered structures; more specifically, as combinatorial geometries - geometric lattices [1]. However, it is necessary to point out that the abovementioned works are dedicated to the introduction into the general academic theory of matroids and therefore they do not present sufficient results for the study of matroids and their mappings specifically as elements of the respective category. More sufficient information on the matter is provided in the articles of such authors as Crapo [2], Nguyen [10, 11], Kelly and Kennedy [7], Lucas [9], Revyakin [22, 23] and others.

This work consists of four parts. The first part contains reference information on the core notions and definitions that characterize matroids as axiomatically defined algebraic objects.

The properties of strong and weak mappings of matroids are described and results are presented that allow matroids to be viewed as objects of the category that are connected by morphisms, sufficient for the description and study of the properties of pseudo-matroids generated by them.

The second part is entirely dedicated to pseudo-matroids generated by elementary mappings of matroids, and the special case of semi-matroids generated by some G-mappings of binary matroids. The terms "pseudo-matroid", along with "cryptomorphic" versions [3], [23] of matroids are used in the theory of matroids for the definition of structures of matroid type. A characteristic feature of pseudo-matroids introduced in this work is that they uniquely define the elementary mapping by which they are generated, and, therefore, there is a functional connection between them and the category of matroids and their mappings. Owing to this fact, an array of general theoretical results has been obtained, which allows for pseudo-matroids to be generated by elementary mappings, introduced in this work, to be viewed as a separate notion in the general theory of matroids.

However, it is important to highlight that in general the family of pseudo-matroids generated by the elementary mappings of an arbitrary matroid is not an axiomatically defined algebraic structure. At the same time, it is proven that the category of binary matroids is closed in respect to morphisms - G-mappings and weak mappings that preserve the rank. This fact allows construction of a family of semi-matroids which is defined by the set of binary matroids and the proof that it coincides with the set of semi-matroids generated by the G-mappings of binary matroids.

The third part of this work is dedicated to one of the unsolved problems of the theory of matroids - the enumeration of all the non-isomorphic matroids defined on a finite set S. It is known that their amount is estimated by a value of $2^{\frac{1}{|S|}2^{|S|}}$. The problem is that to calculate the number of non-isomorphic matroids' non-enumerative ("constructive") algorithms for their extraction (enumeration) from the set of all matroids, the cardinality of which is $|S|!2^{\frac{1}{|S|}2^{|S|}}$, are required, which do not yet exist. At the moment of this work only the results of the enumeration of all non-isomorphic matroids for $1 \le |S| \le 8$ are known.

Defining a matroid means describing the uniquely defining properties of the families of subsets S; for example, its families of cycles. In the current study a matroid is defined mainly as such a family of cycles. Thus, the problem of enumeration comes down to the problem of construction of the families of cycles of all non-isomorphic matroids.

In the first section of the third part the general problem is viewed through the isomorphism of pseudo-matroids generated by the elementary mappings of matroids. An algorithm for the constructive check of isomorphism for arbitrary matroids is defined by the families of their cycles.

The second section is dedicated to binary matroids and respectively to isomorphisms of semi-matroids generated by the G-mappings of binary matroids. The main result is a method of enumeration of all non-isomorphic binary matroids that has exponential complexity. An algorithm for the construction of families of cycles of all non-isomorphic binary matroids with computation complexity of $O(|S|^3 3^{|S|})$ arithmetic operations is presented. Respective results are presented for $1 \le |S| \le 15$. It is worth pointing out, though, that (as opposed to the general case) the number of binary matroids is known precisely; however, the problem of the numeration of non-isomorphic binary matroids for $|S| > 8$ remains unsolved.

In the fourth part of this work the obtained results are used as an instrument in solving some certain practical problems. Using the algorithm from part III, all of the non-isomorphic optimal linear codes for quite a broad range of values of $|S|$ are built. A matroid interpretation for the decoding of the classical problem of linear codes is studied. A new way of coding (G-coding) is described, which generalizes traditional convolutional coding. The algebraic conditions of maximum noise-resistance for G-codes are substantiated. An algorithm for the construction of *G*-codes that fulfill certain conditions, based on the use of non-isomorphic optimal linear codes, is offered. It is shown that an arbitrary convolutional code can be presented by a *G*-code, and it is proven that such *G*-codes do not fulfill the conditions of maximum noise-resistance.

Acknowledgments

The authors express their gratitude to A.A. Nechaev, Professor, PhD, B.A. Pogorelov, Professor, PhD, A.M. Revyakin, Professor, PhD and A.O. Grechkin, PhD, for their methodological assistance in the creation of this work and its preparation for publication.

Chapter 1

INTRODUCTION TO THE THEORY OF MATROIDS

The theory of matroids peaked in development in the middle of the previous century as a result of a general theoretical approach to studying the properties of linear dependences that appear between vectors in vector spaces over fields, and algebraic dependences between the elements of augmentations over fields. A matroid is defined on a finite set and is a population of axiomatically defined dependences of a certain type between the elements of this set.

Suppose $|S|$ is the cardinality of a finite set *S*, while 2^S is the set of all its subsets. For any family of subsets $\mathfrak{A} \subseteq 2^S$, let us define $\hat{\mathfrak{A}}$ as the set of its element, arranged by inclusion.

According to [1], an arranged set $\hat{\mathfrak{A}}$ would be a lattice, if for any of its elements $A, B \in \hat{\mathfrak{A}}$ there exists a unique, minimal by inclusion element $A \vee B$ which is the supremum, such that $A \subseteq A \vee B$ and $B \subseteq A \vee B$, and a single element $A \wedge B$, maximal by inclusion, is the infimum, for which $A \wedge B \subseteq A$ and $A \wedge B \subseteq B$. If $A \subseteq B$, then the set $[A, B] = \left\{C \in \hat{\mathfrak{A}} \,\middle|\, A \subseteq C \subseteq B\right\}$ is called an interval, and if $[A, B] = \{A, B\}$, then the element *B* *covers* the element A in $\hat{\mathfrak{A}}$. If in any interval $[A, B]$ all the maximal covering sequences of the type $A \subseteq C_1 \subseteq C_2 \subseteq ... \subseteq C_k \subseteq B$ have the same length, then the arranged set $\hat{\mathfrak{A}}$ fulfills the sequential condition of Jordan-Dedekind. A lattice $\hat{\mathfrak{A}}$, which fulfills the Jordan-Dedekind condition, is called semi-modular, if for any two elements $A, B \in \hat{\mathfrak{A}}$ from the condition that *A* and *B* covers the infimum $A \wedge B$, it can be inferred that also the supremum $A \vee B$ covers *A* and *B*. A lattice is called geometric if it is semi-modular and each element $A \in \hat{\mathfrak{A}}$ is the supremum of all elements of $\hat{\mathfrak{A}}$, minimal by inclusion, that belong to *A*.

Let us call $\mathbb{B}(S) = 2^S$ the lattice of all subsets of set *S*, arranged by inclusion.

Suppose a set of subsets $\mathfrak{A} \subseteq 2^S$ is given, such that $A \not\subseteq B$ for all $A, B \in \mathfrak{A}$. A family of subsets $J(\mathfrak{A}) = \bigcup_{B \in \mathfrak{A}} \{C \subseteq S \mid C \subseteq B\}$ is called an *ideal*, and a family of subsets $\Phi(\mathfrak{A}) = \bigcup_{B \in \mathfrak{A}} \{C \subseteq S \mid B \subseteq C\}$ is called a *filter*, in the set 2^S.

Respectively arranged sets $J(\mathfrak{A})$ and $\Phi(\mathfrak{A})$ are called an *order ideal* and an *order filter* in lattice $\mathbb{B}(S)$. Let us note that order ideals and filters are uniquely defined by the subsets of the family $\mathfrak{A}$ as its maximal and minimal elements, respectively.

In the case of a formalized definition of families of subsets, the minimality and maximality by inclusion condition of sets $A \subseteq S$, which belong to these families, we shall henceforth denote as $A - \min$ and $A - \max$.

The first part of this work consists of three section and contains the information necessary for the representation of matroids as connected by morphisms – strong and weak mappings - axiomatically defined objects of the category, and sufficient for the description and study of the properties of the pseudo-matroid, introduced in part II, generated by the elementary mappings of matroids. For more detailed acquaintance with the general theory of matroids we refer the reader to the monographs of Welsh [14] and Oxley [12].

1. MAIN NOTIONS AND DEFINITIONS

1.1. Axiomatization of Matroids

A matroid can be described in several equivalent ways. It is said that a matroid is defined on set S, if at the same time it can be said about any subset of this set that it is "closed" or "non-closed", "dependent" or "independent", that it is a "base" or a "cycle", which its "rank" and place in the lattice of closed subsets, arranged by inclusion, are for example: whether it is a "co-point" or not.

The definition [3] is known as the classic definition of a matroid, which utilizes a closure operator $A \to \overline{A},\ A \subseteq S$, that fulfills the exchange condition.

Definition 1. Let us call the mapping $\mu : 2^S \to 2^S$ a closure operator with the property of exchange and label $\mu(A) = \overline{A},\ A \subseteq S$, if for any subsets $A, B \subseteq S$ and elements $a, b \in S$ the following conditions are fulfilled:

1. $A \subseteq \overline{A}$;
2. $A \subseteq \overline{B} \Rightarrow \overline{A} \subseteq \overline{B}$;
3. $a \notin \overline{A},\ a \in \overline{A \cup b} \Rightarrow b \in \overline{A \cup a}$.

In this connection, the set $\overline{A}$ is called a closure of the set A, and a set that coincides with its closure is called a closed set.

Definition 2. A matroid, defined on a finite set S, is an ordered couple $< S, \mathfrak{I} >$, where $\mathfrak{I} = \left\{ \overline{A} \middle| A \subseteq S \right\}$.

From 1) and 2) it can be inferred that the closure operator preserves the order, i.e., $A \subseteq B \Rightarrow \overline{A} \subseteq \overline{B}$, and it is an idempotent: $\overline{A} = \overline{\overline{A}}$ for any subset $A \subseteq S$. The property of exchange 3) for the closure operator is called the axiom of closure for matroids.

Let us denote by $\mathfrak{M}(S)$ a set of matroids defined on the set S. From now on, unless otherwise specifically stated, we shall assume that the matroids considered belong to this set.

For a matroid M a closed set is called its flat, a family of which we shall denote by $\mathfrak{I}_M$. For any subset $A \subseteq S$ the respective flat $\overline{A} \in \mathfrak{I}_M$ we will also denote by $\mathfrak{I}_M(A)$.

It is possible to show [14] that the set of flats $\mathfrak{I}_M$, ordered by inclusion, is a geometric lattice. In the lattice $\mathfrak{I}_M$ the supremum and the infimum for every couple of flats $\overline{A}, \overline{B} \in \mathfrak{I}_M$ are defined by the equations $\overline{A} \vee \overline{B} = \overline{\overline{A} \cup \overline{B}}$ and $\overline{A} \wedge \overline{B} = \overline{A} \cap \overline{B}$. The closure of an empty set $\overline{\varnothing}$ and the set S itself are obviously flats and are, respectively, the minimal and the maximal elements in the lattice $\mathfrak{I}_M$.

Henceforth, for any subsets $A, B \subseteq S$ we shall use the following notation: $A - B = \{a \in A | a \notin B\}$.

Definition 3. An arbitrary subset $A \subseteq S$ is called independent if $a \notin \overline{A - a}$ for all elements $a \in A$. In the contrary case, a subset A is called dependent. An empty set is independent by definition.

The family of all independent and dependent sets of a matroid M we shall respectively denote by $\mathfrak{F}_M$ and $\mathfrak{E}_M$. By means of independent and dependent sets, it is also possible to unequivocally define a matroid M as an ordered couple $< S, \mathfrak{F}_M >$ or $< S, \mathfrak{E}_M >$.

It is easy to show [14] that the family of independent set $\mathfrak{F}_M$ is an ideal in the set 2^S and that the order ideal $\mathfrak{F}_M = J(\mathfrak{B}_M)$ is unequivocally defined by the set of all its maximal elements $\mathfrak{B}_M$, which are called the bases of the matorid M. Analogously, the family of dependent sets $\mathfrak{E}_M$ is a filter in the set 2^S, and the order filter $\mathfrak{E}_M = \Phi(\mathfrak{R}_M)$ is unequivocally defined by the set of its minimal elements $\mathfrak{R}_M$, which are called the cycles of the matroid M.

From this it follows that, having defined matroid M as an ordered couple $< S, \mathfrak{B}_M >$ or $< S, \mathfrak{R}_M >$, i.e., having built the families of the bases and cycles of the matroid, we will then obtain all of its independent and dependents sets. Taking account of this fact, we shall present the axiomatization of matroids through their sets of bases and cycles below.

The family $\mathfrak{B}$ of subsets of the set S is a family of bases of a matroid from the set $\mathfrak{M}(S)$, if the following conditions are fulfilled:

1. $A, B \in \mathfrak{B} \Rightarrow A \not\subseteq B$;
2. $A, B \in \mathfrak{B}, a \in A \Rightarrow \exists b \in B$ and $(A-a) \cup b \in \mathfrak{B}$.

Also, as above, the condition 2) we shall call the axiom of bases for matroids. It follows from this axiom that $|B_1| = |B_2|$ for any $B_1, B_2 \in \mathfrak{B}$.

Let us note that the axiom of bases is known in linear algebra as the replacement theorem.

The family $\mathfrak{R}$ of subsets of the set S is a family of cycles of a matroid M from the set $\mathfrak{M}(S)$, if the following conditions are fulfilled:

1. $A, B \in \mathfrak{R} \Rightarrow A \not\subseteq B$;
2. $A, B \in \mathfrak{R}, a \in A \cap B \Rightarrow \exists\, C \in \mathfrak{R}$ and $C \subseteq A \cup B - a$.

The condition 2) is called the axiom of the cycle for matroids.

Cycles, by definition, are dependent sets, minimal by inclusion, and by their means the closure operator can be easily defined. Indeed, we will (statement 21) later show that the family of subsets $\mathfrak{I}_M = \{\overline{A} \mid A \subseteq S\} \subseteq 2^S$, such that $\overline{A} = \{a \in S \mid a \in A$ or $a \in C \subseteq A \cup a$, $C \in \mathfrak{R}_M\}$ is the family of flats of a matroid M, for which the set $\mathfrak{R}_M$ is the family of its cycles. Therefore, if $a \notin A$ and $a \in \overline{A}$, then there exists a cycle $C \in \mathfrak{R}_M$, such that $a \in C \subseteq A \cup a$. This, in particular, implies that the closure of any base $B \in \mathfrak{B}_M$ would be the whole set S. Any subset $A \subseteq S$ is independent, i.e., $A \in \mathfrak{F}_M$, if and only if it does not contain cycles from the family $\mathfrak{R}_M$. In the contrary case, the subset A would be dependent. In other words, if the matroid $M \in \mathfrak{M}(S)$ is defined as an ordered couple, $< S, \mathfrak{R}_M >$, then exactly this way the families of all dependent and independent sets $\mathfrak{F}_M$ and $\mathfrak{E}_M$ can be built. Particularly, the family of bases $\mathfrak{B}_M$ is defined through the family of cycles $\mathfrak{R}_M$ as follows: $\mathfrak{B}_M = \{B \subseteq S \mid C \not\subseteq B, C \in \mathfrak{R}_M$ and B-$\max\}$.

Henceforth, unless otherwise stated, matroids $M \in \mathfrak{M}(S)$ are set exactly as ordered couples $< S, \mathfrak{R}_M >$. Respectively, the families of flats $\mathfrak{I}_M$ and bases $\mathfrak{B}_M$ are defined in the abovementioned way. Highlighting this fact, let us call the families $\mathfrak{F}_M = \{A \subseteq S \,|\, C \not\subseteq A, C \in \mathfrak{R}_M\}$ and $\mathfrak{E}_M = \{D \subseteq S \,|\, C \subseteq D, C \in \mathfrak{R}_M\}$ also the families of $\mathfrak{R}_M$ - independent - and $\mathfrak{R}_M$ - dependent - subsets of the matroid M .

Let us call any subset $B \subseteq A$ a generating subset *of* set A in matroid M, if $\overline{B} = \overline{A}$. It is easy to see that the bases $\mathfrak{B}_M$ will be the minimal elements of the order filter of all the generating subsets of S, arranged by inclusion.

A matroid $B(S)$ is called a *free matroid*, defined on the set S, if $\mathfrak{I}_{B(S)} = \mathfrak{F}_{B(S)} = \mathbb{B}(S)$. It is obvious that $\mathfrak{B}_{B(S)} = S$ and $\mathfrak{R}_{B(S)} = \varnothing$.

Definition 4. The number $r_M(A) = \max\{|D| \,|\, D \subseteq A, D \in \mathfrak{F}_M\}$ is called the rank of the subset $A \subseteq S$ in matroid M .

It is obvious that the rank of a matroid $r_M(S) = |B|$, where B is any base from the family $\mathfrak{B}_M$. The ranking function r_M unequivocally defines the matroid M and fulfills the *semi-modularity property*, which means that $r_M(A \cup B) + r_M(A \cap B) \le r_M(A) + r_M(B)$ for any subsets $A, B \subseteq S$. Couples of subsets $A, B \subseteq S$, for which in the last inequation equality is achieved, are called modular couples.

It is possible to show [14] that the flat $\overline{A} \in \mathfrak{I}_M$, defined through the ranking function as follows: $\overline{A} = \{a \in S \,|\, r_M(A \cup a) = r_M(A)\}$. Here it is obvious that $r_M(A) = r_M(\overline{A})$.

Definition 5. The flats $\overline{A} \in \mathfrak{I}_M$ are called *co-points* if their rank $r_M(\overline{A}) = r_M(S) - 1$.

If we denote the family of co-points by $\mathfrak{K}_M$, then the ordered couple $< S, \mathfrak{K}_M >$ also unequivocally defines the matroid M.

A family $\mathfrak{K}$ of subsets of the set S is a family of co-points of a matroid from the set $\mathfrak{M}(S)$ if the following conditions are fulfilled:

1. $A, B \in \mathfrak{K} \Rightarrow A \not\subseteq B$;
2. $A, B \in \mathfrak{K}, a \notin A \cup B \Rightarrow \exists K \in \mathfrak{K}$ and $(A \cap B) \cup a \subseteq K$.

We shall call the condition 2) an *axiom of co-points* for matroids.

For a family of co-points the following statement is valid.

Statement 1[14]. *For any flat* $\overline{A} \in \mathfrak{I}_M$ *of rank* $r_M(\overline{A}) = r_M(S) - k$, $k \geq 1$, *there would be a family of co-points* $K_i \in \mathfrak{K}_M, i = \overline{1,k}$, *such that* $\overline{A} = \bigcap_{i=1}^{k} K_i$.

Definition 6. Matroids $M, H \in \mathfrak{M}(S)$ are called *isomorphic* if there exists a one-to-one correspondence π over the set S, such that $A \in \mathfrak{F}_M$, $A \subseteq S$, if and only if $\pi(A) \in \mathfrak{F}_H$.

Families of independent sets can be replaced by families of dependent sets or bases, cycles, flats or co-points in this definition.

1.2. Duality of Matroids

Definition 7. For any matroid M *a dual* matroid would be a matroid M^*, for which $\mathfrak{B}_{M^*} = \{S - B \mid B \in \mathfrak{B}_M\}$. It is obvious that $(M^*)^* = M$.

The following statement establishes the connection between some of the notions introduced earlier for the matroids M and M^*.

Statement 2[14]. *If* $M, M^* \in \mathfrak{M}(S)$ *are dual to one another and* $A \subseteq S$, *then the following equations are fulfilled:*

1. $A \in \mathfrak{F}_{M^*} \Leftrightarrow \mathfrak{I}_M(S - A) = S$;
2. $r_{M^*}(A) = |A| + r_M(S - A) - r_M(S)$;
3. $A \in \mathfrak{R}_{M^*} \Leftrightarrow S - A \in \mathfrak{K}_M$.

As is clear from statement 2, the notions of a dual matroid mentioned there are not necessarily denominated through similar notions for the initial matroid. In this respect, the property 3) is very important, as it connects the cycles and the co-points of matroids and their dual matroids.

Let us present some more properties that instantiate the actual connections between the families of the bases and cycles of matroids and their dual matroids.

Statement 3[14]. *Suppose matroids* $M, M^* \in \mathfrak{M}(S)$ *are dual to one another and* $A \subseteq S$. *Then the following equations are fulfilled:*

1. $\mathfrak{R}_{M^*} = \{A \subseteq S \| A \cap C | \neq 1, C \in \mathfrak{R}_M$ *and* $A - \min\}$;
2. $\mathfrak{R}_{M^*} = \{A \subseteq S | A \cap B \neq \varnothing, B \in \mathfrak{B}_M$ *and* $A - \min\}$.

1.3. Binary Matroids

Suppose $S = \{1, 2, ..., n\}$ and a $k \times n$ matrix M' of rank k is set over the field $GF(2)$, its columns indexed by the elements of the set S. Let us define a matroid $M \in \mathfrak{M}(S)$, such that $A \in \mathfrak{F}_M$, where $A = \{i_1, i_2, ..., i_l\} \subseteq S, l \geq 1$, if and only if the vector-columns of matrix M' with numbers $i_1, i_2, ..., i_l$ are linearly independent. It is obvious that the matroid's rank is $r_M(S) = k$, the bases $\mathfrak{B}_M$ are sets of k vector-columns, to which the sub-matrixes of rank k conform, and so on. Such a matroid is called a vector matroid.

A matroid M is called a binary matroid if it is isomorphic to a vector matroid. Matroids that can be depicted in this way over any field are called regular.

Example 1. The Fano matroid is a binary matroid of rank 3, defined on the set $S, |S| = 7$ and can be depicted by the matrix $\begin{Vmatrix} 1001101 \\ 0101011 \\ 0010111 \end{Vmatrix}$. The Fano matroid is an example of a binary, but not depictable over any other field and, therefore, a non-regular matroid [14].

From now on for any subsets $A, B \subseteq S$ we shall call their symmetric difference a binary sum and denote $A \Delta B = (A - B) + (B - A) = A \oplus B$. Respectively, for the populations of subsets $A_i \subseteq S, i = \overline{1,k}$, and their symmetric difference $A_1 \Delta A_2 \Delta \ldots \Delta A_k = \sum_{i=1}^{k} \oplus A_i$.

To respectively denote the direct sum and to join an unspecified number of subsets of the set S we shall use the expressions "$\sum A, A \subseteq S$" and "$\bigcup A, A \subseteq S$".

Statement 4[14]. *A matroid* M *is binary if and only if any of the following conditions is fulfilled:*

1. $D_1, D_2 \in \mathfrak{R}_M \Rightarrow D_1 \oplus D_2 = \sum C, C \in \mathfrak{R}_M$;

2. $D \in \mathfrak{R}_M, D^* \in \mathfrak{R}_{M^*} \Rightarrow |D \cap D^*|$ *is even.*

The condition 1) of statement 4 we shall later refer to as the axiom of cycles for binary matroids.

Example 2. Suppose $|S| = n$ and $M(n,k)$ is a matroid of rank k, the family of bases of which is $\mathfrak{B}_{M(n,k)} = \{B \subseteq S \| B| = k\}$. Such a matroid is called uniform [14]. It is obvious that $\mathfrak{R}_{M(n,k)} = \{C \subseteq S \| C| = k + 1\}$ is a family of cycles of the uniform matroid $M(n,k)$. Therefore, $M(n,n) = B(S)$, so that $\mathfrak{B}_{M(n,n)} = S$ and $\mathfrak{R}_{M(n,n)} = \varnothing$. It is also clear that $M^*(n,k) = M(n,n-k)$. As can be easily seen, the axiom of cycles for binary matroids implies that for $n \geq 4$ and $2 \leq k \leq n-2$ uniform matroids $M(n,k)$ are not binary.□

For any base $B \in \mathfrak{B}_M$ of a matroid M and any element $a \in S - B$ there exists a unique cycle $C(a,B) \in \mathfrak{R}_M$, such that $a \in C(a,B) \subseteq B \cup a$. Cycle $C(a,B)$ is called a *fundamental cycle of the element* $a \in S - B$ in the base $B \in \mathfrak{B}_M$.

It is easy to show that for binary matroids fundamental cycles fulfill the following property: for any arbitrary cycle $C \in \mathfrak{R}_M$, if $C - B = \{a_1, \ldots, a_t\}$, an equality is fulfilled:

$$C = C(a_1, B) \oplus \ldots \oplus C(a_t, B). \tag{1}$$

For any matroid M the element $a \in S$, for which $r_M(a) = 0$, is called a *loop*. Let us note that in these terms the closure of an empty set is set by the formula $\overline{\varnothing} = \{a \in S \mid r_M(a) = 0\}$. For example, in matroid $B^*(S)$ all the elements of the set S are loops. Similarly, all the elements $a,b \in S$, for which $r_M(\{a,b\}) = 1$, are called *parallel*. A matroid M without loops and parallel elements is called a *simple matroid*.

Let us note that the equation $C(a,B) = \{a\}$ is fulfilled for any loop $a \in S$ and any base $B \in \mathfrak{B}_M$.

1.4. Sub-Matroids

Let us recall that a sub-lattice of any lattice is a subset of its elements along with their supremums and infimums [1].

Definition 8. A matroid H is called *sub-matroid* of matroid M, if the lattice $\mathfrak{I}_H$ is a sub-lattice of the lattice $\mathfrak{I}_M$.

In the theory of matroids two methods for the construction of sub-matroids are commonly used, specifically the procedures of “restriction” and “contraction”. A sub-matroid generated by a composition of restrictions and contractions is called a minor of the initial matroid.

Definition 9. A *restriction* of a matroid $M \in \mathfrak{M}(S)$ on the set $A \subseteq S$ is a sub-matroid $M|A$ with the family of flats of the following kind:

$$\mathfrak{I}_{M|A}(B) = \mathfrak{I}_M(B) \cap A \text{ for all subsets } B \subseteq A.$$

In the lattice $\mathfrak{I}_M$ a sub-lattice $[\overline{\varnothing}, \mathfrak{I}_M(A)]$ corresponds with the sub-matroid $M \mid A$. Sub-matroids, which are generated by the procedure of restriction, have the following properties.

Statement 5[14]. *Suppose* $A \subseteq S$. *Then the following hold true:*

1. $B \in \mathfrak{C}_M, B \subseteq A \Rightarrow B \in \mathfrak{C}_{M|A}$;
2. $r_{M|A}(A) = r_M(A)$;
3. $C \in \mathfrak{R}_M, C \subseteq A \Rightarrow C \in \mathfrak{R}_{M|A}$;
4. $\mathfrak{K}_{M|A} = \{H \cap A \mid H \in \mathfrak{K}_M, H \cap A \neq A \text{ and } H \cap A - \max\}$.

Definition 10. A *contraction* of a matroid $M \in \mathfrak{M}(S)$ on the subset $A \subseteq S$ is a sub-matroid $M \bullet A$ with the family of flats of the following kind

$\mathfrak{I}_{M \bullet A}(B) = \mathfrak{I}_M((S - A) \cup B) - (S - A)$ for all subsets $B \subseteq A$.

The lattice of flats of a sub–matroid $M \bullet A$ is isomorphic to the sub–lattice $[\mathfrak{I}_M(S - A), S]$ in the lattice $\mathfrak{I}_M$.

Sub-matroids generated by the contraction of matroids fulfill the following properties.

Statement 6[14]. *Suppose* $A \subseteq S$. *Then the following hold true:*

1. $r_{M \bullet A}(B) = r_M((S - A) \cup B) - r_M(S - A)$, $B \subseteq A$;

2. $\mathfrak{R}_{M \bullet A} = \{C \cap A | C \in \mathfrak{R}_M, C \cap A \neq \varnothing$ *and* $C \cap A$ $-\min$ };
3. $\mathfrak{I}_{M \bullet A}(B) = A \Leftrightarrow \mathfrak{I}_M((S - A) \cup B) = S$.

The operations of restriction and contraction for dual matroids are defined in the same way and are inter-connected.

Statement 7. *Suppose* $A \subseteq S$*. Then the following equations hold true*:

1. $(M \mid A)^* = M^* \bullet A$;
2. $(M \bullet A)^* = M^* \mid A$.

Let us list a few results for multiple restrictions and contractions.

Statement 8. *Suppose* $A \subseteq S$ *and* $B \subseteq A$*. Then the following equations hold true*:

1. $M \mid B = (M \mid A) \mid B$;
2. $M \bullet B = (M \bullet A) \bullet B$;
3. $(M \mid A) \bullet B = (M \bullet (S - (A - B))) \mid B$;
4. $(M \bullet A) \mid B = (M \mid (S - (A - B))) \bullet B$.

1.5. Connectivity of Matroids

Definition 11. A *bridge* of any subset $A \subseteq S$ in a matroid $M \in \mathfrak{M}(S)$ is an element $a \in A$, such that $r_M(A - a) < r_M(A)$.

It follows from the definition that the subset A has a bridge if and only if there is an element $a \in A$ that belongs to all the sets $B \subset A$, such that $r_M(B) = r_M(A)$. If $A = S$, then the bridge is an element that belongs to all the bases of the matroid. Let us note that the bridge of a matroid would be a loop in the respective dual matroid.

In graph theory a bridge is an edge, the cancellation of which leads to the graph's disconnection. Analogous to graphs, the notions of "connectivity" and "connectivity component" also exist for matroids. Clearly, any connectivity component of a matroid must be its sub-matroid, meaning that the respective lattice of flats must be a sub-lattice of the lattice $\mathfrak{I}_M$.

Definition 12. A matroid $M \in \mathfrak{M}(S)$ is called *connected*, or inseparable, if for any elements $a, b \in S$ there exists a cycle $C \in \mathfrak{R}_M$, such that $a, b \in C$.

Therefore, if for any two elements $a, b \in S$ the equivalence $\sim$ relation is introduced as follows:

$$a \sim b \Leftrightarrow \exists\, C \in \mathfrak{R}_M \text{ and } a, b \in C,$$

then the respective equivalence classes of this relation define the *connectivity components* of matroid M .

It follows from definition 12 that the subset $A \subseteq S$ would be a connectivity component of matroid M if and only if the sub-matroid $M \mid A$ is connected.

Hence, a matroid M is *not connected*, or *separable*, if and only if there exists a connectivity component $A \subset S$.

The following statement contains the separability criterion for a matroid.

Statement 9[14]. *A matroid $M \in \mathfrak{M}(S)$ is separable if and only if there exists a subset $A \subset S$, such that $r_M(A) + r_M(S - A) = r_M(S)$.*

In the theory of matroids, especially in the description of matroid mappings, one of the key notions is the notion of a “cyclic flat”.

Definition 13. A flat $\overline{A} \in \mathfrak{I}_M$ is called *cyclic* if it does not contain bridges.

It follows from the definition that a flat $\overline{A} \in \mathfrak{I}_M$ would be cyclic if and only if $a \in \overline{\overline{A} - a}$ for all elements $a \in \overline{A}$. Hence, a cyclic flat can be defined alternatively as a flat that equals a union of cycles.

Let us note that the closure of any cycle $C \in \mathfrak{R}_M$ is a cyclic flat. Indeed, if there is an element $a \notin C$ and $a \in \overline{C}$, then this means that there exists a cycle $D \in \mathfrak{R}_M$, such that $a \in D \subseteq C \cup a$, i.e., $a \in C \cup D$.

As a conclusion of the first section of part I, let us highlight that it contains reference information on the basic notions and definitions that characterize matroids as axiomatically defined algebraic objects. Apart from that, it also contains the results that are necessary for the study of the category of matroids and their mappings. More profound and systemized information on the general properties of matroids, including the proofs of the statements that have been left out from the text, can be found in the monographs of Welsh [14] and Oxley [12].

2. Strong Mappings of Matroids

Matroids from the set $\mathfrak{M}(S)$ are defined on the same set S. Therefore, an identity mapping on S generates set-theory connections of families of sets that unequivocally define a matroid. Particularly, an identity mapping canonically extends to a mapping of the respective lattices of the matroid`s flats. This fact enables us to introduce mappings between matroids and to view them as objects of the category, connected by morphisms.

2.1. Modular Cuts and Filters

Definition 14. Matroids $M, H \in \mathfrak{M}(S)$ are connected by a strong mapping $\varphi_S : M \to H$ and the matroid H is called a *factor* of matroid M if $\mathfrak{I}_H \subseteq \mathfrak{I}_M$.

Any strong mapping between non-isomorphic matroids means that $\mathfrak{I}_H \subset \mathfrak{I}_M$ and $r_H(S) \leq r_M(S) - 1$. If $r_H(S) = r_M(S) - 1$, then such strong mapping is called *elementary* (from here on - simply an *elementary mapping*), and the respective factor is called an *elementary factor*.

In the following statement alternative definitions for strong mappings are given.

Statement 10[14]. *For matroids $M, H \in \mathfrak{M}(S)$ the following conditions are equivalent:*

1. *the mapping $\varphi_S : M \to H$ is strong*;
2. $\mathfrak{I}_M(A) \subseteq \mathfrak{I}_H(A)$, $A \subseteq S$;
3. $B \subseteq A \subseteq S \Rightarrow r_M(A) - r_M(B) \geq r_H(A) - r_H(B)$;
4. *the mapping $\varphi_S^* : H^* \to M^*$ is strong.*

Definition 15. A *subset* $\mathfrak{Y}_M \subseteq \mathfrak{I}_M$ is called a *modular cut* of the family of flats $\mathfrak{I}_M$ of matroid M, if:

1. $\mathfrak{Y}_M$ is an order filter in lattice $\mathfrak{I}_M$;
2. $\overline{A}, \overline{B} \in \mathfrak{Y}_M$, $\overline{A}, \overline{B}$ is a modular couple $\Rightarrow \overline{A} \cap \overline{B} \in \mathfrak{Y}_M$.

Let us note that if the flats $\overline{A}$ and $\overline{B}$ are a modular pair, then the subsets $A, B \subseteq S$ themselves would not necessarily be a modular pair.

Definition 16. A subset $\Phi_M \subseteq 2^S$ is called a *modular filter* of matroid M if:

1. Φ_M is an order filter of the lattice $\mathbb{B}(S)$;
2. $\Phi_M = \{A \subseteq S \mid \overline{A} \in \mathfrak{Y}_M\}$, where $\mathfrak{Y}_M$ is a modular cut of the family of flats $\mathfrak{I}_M$.

It follows from definition 16 that to any modular cut $\mathfrak{Y}_M$ of the matroid M uniquely corresponds a modular filter $\Phi_M = \{A \subseteq S \mid \overline{A} \in \mathfrak{Y}_M\}$. In turn, any modular filter Φ_M generates a modular cut $\mathfrak{Y}_M = \Phi_M \cap \mathfrak{I}_M$. So, there is one-to-one correspondence between the families of modular cuts and those of modular filters.

A modular filter of the matroid M is called *proper* if it is not an empty set and the respective modular cut $\mathfrak{Y}_M \neq \mathfrak{I}_M$.

There is a one-to-one correspondence between the elementary factors and the proper modular filters of any matroid M [14]: if Φ_M is a proper modular filter of the matroid M, then the function

$$r_H(A) = \begin{cases} r_M(A) - 1, & if\ A \in \Phi_M \\ r_M(A), & if\ A \notin \Phi_M \end{cases} \tag{2}$$

will be the rank function of elementary factor H of the matroid M. If H is an elementary factor of the matroid M, then $\{A \subseteq S \mid r_H(A) = r_M(A) - 1\}$ is a proper modular filter of the matroid M.

Definition 17. A subset $\mathfrak{L}_M \subseteq \mathfrak{K}_M$ is called a *linear cut* of a family of co-points $\mathfrak{K}_M$ if for any $K_1, K_2 \in \mathfrak{L}_M$, that cover the flat $K_1 \cap K_2$ in the lattice $\mathfrak{I}_M$, any co-point $K \in \mathfrak{K}_M$, covering the flat $K_1 \cap K_2$, also belongs to the set $\mathfrak{L}_M$.

In [3] it is shown that there exists a one-to-one correspondence between the dissimilar to S modular cuts of the family of flats $\mathfrak{I}_M$ and the linear cuts of the family of co-points $\mathfrak{K}_M$ of the matroid M. In fact, the co-points that belong to any modular cut form a linear cut of the family of co-points, and vica versa, the co-points of a linear cut would be co-points of some modular cut.

However, a situation in which the modular cut of the family of flats of a matroid coincides with the set S, and it therefore does not contain any co-points from $\mathfrak{K}_M$, is also possible.

2.2. Truncation of Matroids and Erectable Matroids

The situation when a modular cut of a matroid coincides with the set S is matched by a mapping called the "top reduction" or the "truncation" of matroids.

Definition 18. For a matroid $M \in \mathfrak{M}(S)$ its *top reduction (truncation)* is a matroid $T(M)$, the family of independent sets of which is defined as follows:

$$\mathfrak{F}_{T(M)} = \{A \subseteq S \,|\, A \in \mathfrak{F}_M \; and \; |A| < r_M(S)\}. \tag{3}$$

It is obvious that $r_{T(M)}(S) = r_M(S) - 1$ and all the bases $B \in \mathfrak{B}_M$ of the matroid $T(M)$ are dependent sets, i.e., $B \in \mathfrak{C}_{T(M)}$. Moreover, below we will show that $B \in \mathfrak{R}_{T(M)}$.

A top reduction of the lattice $\mathfrak{I}_M$ means the deletion of the co-points $\mathfrak{K}_M$ and their replacement with the set S. This implies that, first of all,

$\mathfrak{I}_{T(M)} \subset \mathfrak{I}_M$, and, therefore, a top reduction would be a strong elementary mapping and, second of all, that a top reduction of a matroid is not a sub-matroid.

If the matroid $T(M)$ is the result of a top reduction of the matroid M, then the matroid M, in turn, is called an *erection* of the matroid $T(M)$. Under a trivial erection of a matroid $T(M)$ we refer to the matroid $T(M)$ itself. Therefore, for the respective lattices of flats, an erection is the addition of a new level of co-points to the lattice $\mathfrak{I}_{T(M)}$, so that the new lattice would be a $\mathfrak{I}_M$ lattice of some matroid M. Matroids that allow non-trivial erection are called *erectable*.

Definition 19. For any $k, k \geq 1$, a subset $A \subseteq S$ we shall call k-closed in the matroid $M \in \mathfrak{M}(S)$, if $\overline{B} \subseteq A$ for any $B \subseteq A$, such that $|B| \leq k$. In turn, a k*-closure* of a set A is the k-closed set, minimal by inclusion, that contains A.

As an intersection of k-closed sets is a k-closed set, the definition is correct.

Statement 11. *A subset* $A \subseteq S$ *would be* k*-closed,* $k \leq |A|$, *in the matroid* $M \in \mathfrak{M}(S)$ *if and only if* $A \cap \overline{B} \in \mathfrak{I}_M$ *for all the flats* $\overline{B} \in \mathfrak{I}_M$, *such that* $|\overline{B}| = k$.

Proof. Suppose that the subset $A \subseteq S$ is k-closed in the matroid M and $\overline{B}$ is a k-flat (i.e., $\overline{B} \in \mathfrak{I}_M$ and $|\overline{B}| = k$). As $A \cap \overline{B} \subseteq \overline{B}$, $r_M(A \cap \overline{B}) \leq k$. Suppose $D \in \mathfrak{F}_M$ is any independent subset of the set $A \cap \overline{B}$, such that $\overline{D} = \overline{A \cap \overline{B}}$. As $|D| \leq k$, then, according to the proposition, $\overline{D} \subseteq A$. In turn, as $\overline{B} \in \mathfrak{I}_M$, then $\overline{D} \subseteq \overline{B}$, so that $\overline{A \cap \overline{B}} = \overline{D} \subseteq A \cap \overline{B}$ and, therefore, the subset $A \cap \overline{B} \subseteq S$ is closed, i.e., $A \cap \overline{B} \in \mathfrak{I}_M$.

Conversely, let us consider the subset $A \subseteq S$ and that $A \cap \overline{B} \in \mathfrak{I}_M$ for all the k-flats $\overline{B} \in \mathfrak{I}_M$. If $C \subseteq A$ and $|C| = k$, then the rank of the flat $\overline{C} \in \mathfrak{I}_M$ is $r_M(\overline{C}) \leq k$ and, according to the proposition, $A \cap \overline{C} \in \mathfrak{I}_M$. The inclusion $C \subseteq A \cap \overline{C} \subseteq \overline{C}$ is valid, and hence $\overline{C}$ is the minimal closed set that contains the subset $C \subseteq A$. Therefore, $A \cap \overline{C} = \overline{C}$, so that $\overline{C} \subseteq A$ and hence the subset $A \subseteq S$ is k-closed. □

The following statement contains the main result regarding the problem of the erectability of matroids.

Statement 12[2]. *Suppose $M \in \mathfrak{M}(S)$ is a simple matroid of rank k. The family $\mathfrak{K}$ of subsets of the set S would be a family of co-points of the erection of the matroid M if and only if the following conditions are fulfilled:*

1. $A \in \mathfrak{K} \Rightarrow \mathfrak{I}_M(A) = S$;
2. $A \in \mathfrak{K} \Rightarrow$ *subset* $A \subseteq S$- $(k-1)$- *closed in* M;
3. *every base of the matroid* M *contained in only one subset of the family* $\mathfrak{K}$.

The flat $\overline{A} \in \mathfrak{I}_M$ is called *essential* if the sub-matroid $M \mid \overline{A}$ is erectable. It is possible to show [2] that the set of essential flats and their ranks unequivocally define a matroid.

2.3. Factorization of Strong Mappings

If there is an elementary mapping between matroids M and H, we shall use the notation $M \xrightarrow{\Phi} H$ or $M \xrightarrow{\mathfrak{Y}} H$ and say that this elementary mapping is generated by the respective proper modular filter Φ_M, or (when speaking of flats) a modular cut $\mathfrak{Y}_M$, dissimilar to the family of flats $\mathfrak{I}_M$.

For an arbitrary strong mapping $\varphi_S : M \to H$ the number $n_{\varphi_S}(A) = r_M(A) - r_H(A)$ we shall call the *order of the subset* $A \subseteq S$.

The *order of a strong mapping* is the order of the set S. If $n_{\varphi_S}(S)=k$ then the respective strong mapping we shall denote by $\varphi_S^{(k)}: M \to H$.

Definition 20. A matroid $N \in \mathfrak{M}(S)$ is called a *lift of matroid* H *in matroid* M if there exists a strong mapping $\varphi_S^{(k)}: M \to H$ of order $k, k \geq 1$, which can be consequently represented by a strong mapping $\varphi_S^{(k-1)}: M \to N$ of order $(k-1)$ and an elementary mapping $N \xrightarrow{\mathfrak{Y}} H$.

For any factor H of the matroid M there are always lifts. In particular, a Higgs lift is a matroid N, for which $\Im_N = \Im_H \cup \{\overline{A} \in \Im_M \mid n_{\varphi_S}(\overline{A}) = 0\}$.

In definition 20 the situation $k=1$ corresponds with the fact that matroid M is a lift in matroid M for any of its elementary factors H by definition. Moreover, it is a Higgs lift.

Indeed, suppose $\overline{A} \in \Im_M - \Im_H$. Then $\overline{A} \subset \Im_H(\overline{A})$ and $\Im_H(\overline{A}) \in \Im_M$, and therefore $r_M(\Im_H(\overline{A})) = r_M(\overline{A}) + 1 \geq r_H(\overline{A}) + 1$. As $r_H(\overline{A}) = r_H(\Im_H(\overline{A}))$, then the latter inequation means that $\Im_H(\overline{A}) \in \mathfrak{Y}_M$ and $r_H(\overline{A}) = r_M(\Im_H(\overline{A})) - 1 = r_M(\overline{A})$.

It follows from the reasoning above that for any flat $\overline{A} \in \Im_M - \Im_H$ there exists a unique flat $\Im_H(\overline{A}) \in \mathfrak{Y}_M$ that covers it in the lattice $\Im_M$.

By continuing the *lifting* procedure, we will obtain the factorization of a strong mapping $\varphi_S^{(k)}: M \to H$ of order k:

$$M = H_0 \xrightarrow{\mathfrak{Y}_1} H_1 \xrightarrow{\mathfrak{Y}_2} \ldots \xrightarrow{\mathfrak{Y}_k} H_k = H, \tag{4}$$

where the mappings $H_{i-1} \xrightarrow{\mathfrak{Y}_i} H_i$ are elementary and generated by modular cuts $\mathfrak{Y}_i$ (or modular filters Φ_i for all $i = \overline{1,k}$, respectively). The factorization

(4) is called *Higgs factorization* if every matroid H_{i-1} is a Higgs lift of the strong mapping $\varphi_S^{(i)} : M \to H_i$, $i = \overline{1,k}$.

Modular cuts $\mathfrak{Y}_i$ as order filters in lattices $\mathfrak{I}_{H_{i-1}}, i = \overline{1,k}$, are generated by minimal flats that are called *generators*. Let us note that every cut can have several generators. If a cut has one generator, it is called the *main* cut.

For any strong mapping $\varphi_S : M \to H$ the order of a set is a non-decreasing function, i.e., if $A \subseteq B$ then $n_{\varphi_S}(A) \leq n_{\varphi_S}(B)$. Hence, and from (2), for any factorization (4) order $n_{\varphi_S}(A)$ equals the number of modular filters Φ_i, $i = \overline{1,k}$, that contain the subset $A \subseteq S$.

Definition 21. For any strong mapping $\varphi_S : M \to H$ the *essential order* of the subset $A \subseteq S$ is the value

$$N_{\varphi_S}(A) = n_{\varphi_S}(\mathfrak{I}_H(A)) - \max\{n_{\varphi_S}(\overline{B}) \mid \overline{B} \in \mathfrak{I}_H, \overline{B} \subset \mathfrak{I}_H(A)\}. \qquad (5)$$

As order n_{φ_S} is a non-decreasing function, the maximum in (5) is achieved in a flat $\overline{B} \in \mathfrak{I}_H$, which is covered by the flat $\mathfrak{I}_H(A) \in \mathfrak{I}_H$ in the lattice $\mathfrak{I}_H$.

2.4. The Higgs Factorization of Canonical Mappings

A strong mapping $B(S) \to M$ for any matroid $M \in \mathfrak{M}(S)$ is called canonical. For this case we shall denote the order of the subset $A \subseteq S$ by $n(A) = |A| - r_M(A)$, and, in turn, the essential order as $N(A) = |\overline{A}| - |\overline{B}| - (r_M(\overline{A}) - r_M(\overline{B}))$, where the flat $\overline{B} \in \mathfrak{I}_M$ is covered by the flat $\overline{A} \in \mathfrak{I}_M$ in the lattice the $\mathfrak{I}_M$.

Statement 13. *Suppose $M \in \mathfrak{M}(S)$ and the sequence of mappings*

$$B(S) = M_0 \xrightarrow{\mathfrak{Y}_1} M_1 \xrightarrow{\mathfrak{Y}_2} \ldots \xrightarrow{\mathfrak{Y}_k} M_k = M \qquad (6)$$

is a Higgs factorization of a canonical mapping $B(S) \to M$. *Then the generators of modular cuts* $\mathfrak{Y}_i$, $i = \overline{1,k}$, *are cyclic flats of the matroid* M, *and their essential order* $N(\overline{A})$ *is the number of modular cuts* $\mathfrak{Y}_i$, $i = \overline{1,k}$, *for which the flat* $\overline{A} \in \mathfrak{J}_M$ *is the generator.*

Proof. Suppose the flat $\overline{A} \in \mathfrak{J}_M$ has a bridge, i.e., there exists an element $a \in \overline{A}$, such that $r_M(\overline{A} - a) = r_M(\overline{A}) - 1$. Therefore, $a \notin \overline{\overline{A} - a}$ and the flat $\overline{B} = \overline{\overline{A} - a} \subset \overline{A}$ is covered by the flat $\overline{A}$ in the lattice $\mathfrak{J}_M$. Hence, for the canonical mapping $B(S) \to M$ we will get

$$N(\overline{A}) = |\overline{A}| - |\overline{B}| - (r_M(\overline{A}) - r_M(\overline{B})) = 0.$$

Conversely, if for some flat $\overline{A} \in \mathfrak{J}_M$ under a canonical mapping of $B(S) \to M$ essential order is zero, then $|\overline{A}| - |\overline{B}| = r_M(\overline{A}) - r_M(\overline{B}) = 1$, as the flat $\overline{A}$ covers the flat $\overline{B}$ in the lattice $\mathfrak{J}_M$. So, there exists an element $a \in \overline{A}$, such that $r_M(\overline{A} - a) = r_M(\overline{B}) < r_M(\overline{A})$, and, according to definition 11, the element a is a bridge in the set $\overline{A}$.

To conclude, a flat of any matroid M is cyclic if and only if its essential order under a canonical mapping of $B(S) \to M$ is not zero.

Suppose a matroid M_{k-1} is a Higgs lift of a canonical mapping $B(S) \to M$, i.e., $B(S) \longrightarrow M_{k-1} \xrightarrow{\mathfrak{Y}_k} M$ and $\mathfrak{Y}_k$ is a modular cut for which all the flats of a matroid M_{k-1} are, in fact, flats from the family $\mathfrak{J}_M$, and flats $\overline{B} \in \mathfrak{J}_{B(S)}$ are such that $r_{B(S)}(\overline{B}) = |\overline{B}| = r_M(\overline{B})$. If now the flat $\overline{A} \in \mathfrak{J}_M$ is the minimal by inclusion or a generator of the modular cut $\mathfrak{Y}_k$, then for any flat

$\overline{B} \subset \overline{A}$ which is covered by the flat $\overline{A}$ in the lattice $\Im_M$, firstly, $r_M(\overline{A}) = r_M(\overline{B})$, and secondly, $|\overline{B}| < |\overline{A}|$, i.e., its essential order $N(\overline{A})$ is not zero, and, therefore, as has been shown above, it is cyclic. Then, as $r_M(\overline{B}) = r_{M_{k-1}}(\overline{B})$ and $r_M(\overline{A}) = r_{M_{k-1}}(\overline{A}) - 1$, the $|\overline{A}| - |\overline{B}| - (r_M(\overline{A}) - r_M(\overline{B})) - [|\overline{A}| - |\overline{B}| - (r_{M_{k-1}}(\overline{A}) - r_{M_{k-1}}(\overline{B})] = 1$, and the essential order of the flat $\overline{A}$ under a canonical mapping of $B(S) \to M_{k-1}$ is less than the essential order $N(\overline{A})$ by 1. Taking into account $\overline{A} \in \Im_{M_{k-1}}$ and continuing the Higgs factorization for the canonical mapping $B(S) \longrightarrow M_{k-1}$ and so on in the same fashion, at the end we shall obtain that all the generators of modular cut $\mathfrak{Y}_i, i = \overline{1,k}$, first of all, are cyclic flats $\overline{A} \in \Im_M$ and, second of all, each of them is a generator in factorization (6) $N(\overline{A})$ times.□

The essential flats of any matroid M are cyclic. Indeed, otherwise the flat $\overline{A} \in \Im_M$ would have a bridge $a \in \overline{A}$, that would belong to all the bases of the sub-matroid $M|\overline{A}$, and, therefore, to all the generating subsets of the set $\overline{A}$ in it. If the sub-matroid $M|\overline{A}$, is erectable, then, according to statement 12, all the newly added co-points must contain the element $a \in \overline{A}$. This, along with statement 1, implies that the element $a \in \overline{A}$ must be a loop in the sub-matroid $M|\overline{A}$, i.e., must not belong to any of the bases, which is impossible.

The essential flats of a matroid M as cyclic flat can be generators under a canonical mapping of $B(S) \to M$. There are matroids called the *main matroids*, for which only essential flats are such generators.

Let us present some results by Kelly and Kennedy [6], that are related to the Higgs factorization of strong mappings.

Statement 14. *Suppose:* $\varphi_S^{(k)} : M \to H$ *is a strong mapping of order* k. *Then the following holds true:*

1. *the factorization (6) is a Higgs factorization if and only if* $\mathfrak{Y}_1 \subseteq \mathfrak{Y}_2 \subseteq \ldots \subseteq \mathfrak{Y}_k$;

2. *the factorization (6) is a Higgs factorization if and only if* $H^* = H_k^* \longrightarrow H_{k-1}^* \longrightarrow \ldots \longrightarrow H_0^* = M^*$ *– is a Higgs factorization of a strong mapping* $\varphi_S^{*(k)} : H^* \to M^*$*;*
3. *if (6) is a Higgs factorization of a mapping* $\varphi_S^{(k)}$*, then for any* $i = \overline{1,k}$ *the modular filters are defined as follows:*

$$\Phi_i = \{A \subseteq S \mid n_{\varphi_S^{(i)}}(A) > k - i\}.$$

So, in the second section strong and elementary strong mapping between matroid $M, H \in \mathfrak{M}(S)$ have been defined. An elementary mapping $M \xrightarrow{\Phi} H$ is characterized by a property that it is unequivocally defined by the modular filter Φ_M or the flats of the respective modular cut $\mathfrak{Y}_M$. At the same time, a subfamily of flats $\mathfrak{I}_M - \mathfrak{I}_H$ also has a characteristic property: for any flat $\overline{B} \in \mathfrak{I}_M - \mathfrak{I}_H$ there will always be a single flat $\overline{A} \in \mathfrak{Y}_M$, which covers it in the lattice $\mathfrak{I}_M$. Modular filters and modular cuts as order filters in the respective lattices are unequivocally defined by their minimal elements. It is important, though, that according to [12] and [14], their type is not determined. In part II it will be demonstrated that for any elementary mapping $M \xrightarrow{\Phi} H$ of matroids $M, H \in \mathfrak{M}(S)$ the minimal elements of an order filter Φ_M are unequivocally represented by some cut of the family of cycles $\mathfrak{R}_H$.

3. Weak Mappings of Matroids

3.1. Weak Cuts of Matroids

Definition 22. Matroids $M, H \in \mathfrak{M}(S)$ are connected by a *weak* mapping $\varphi_W : M \to H$, if $\mathfrak{F}_H \subseteq \mathfrak{F}_M$.

It is obvious that for a weak mapping of non-isomorphic matroids $\mathfrak{F}_H \subset \mathfrak{F}_M$. However, as opposed to strong mappings, a case $r_H(S) = r_M(S)$ is also possible.

As with the strong ones, there exist alternative methods of definition for weak mappings.

Statement 15[12]. *For matroids* $M, H \in \mathfrak{M}(S)$ *the following conditions are equivalent:*

1. *the mapping* $\varphi_W : M \to H$ *is weak;*
2. $C \in \mathfrak{R}_M \Rightarrow \exists\ D \in \mathfrak{R}_H$ *and* $D \subseteq C$*;*
3. $r_M(A) \geq r_H(A)$, $A \subseteq S$*;*
4. $r_M(S) = r_H(S) \Rightarrow$ *the mapping* $\varphi_W^* : H^* \to M^*$ *is weak.*

Let us note that the condition 3) of statement 15 is a special case of the condition 3) of statement 10 for $B = \varnothing$, and, therefore, any strong mapping is also a weak mapping.

It follows from definition 22 that under a weak mapping $\varphi_W : M \to H$ the ideal $\mathfrak{F}_M$ would contain the sub-ideal $\mathfrak{F}_H$ of the independent sets of the matroid H and the sub-filter $\mathfrak{F}_M - \mathfrak{F}_H$ of the dependent sets of the matroid H. In particular, if $M = B(S)$ is a free matroid, then the difference $\mathfrak{F}_{B(S)} - \mathfrak{F}_H = 2^S - \mathfrak{F}_H$ is precisely the aggregate of all dependent sets of the matroid H.

Definition 23. A subset of independent sets $\mathfrak{G}_M \subseteq \mathfrak{F}_M$ is called a *weak cut of the matroid* M if the following two conditions are fulfilled:

1. $\mathfrak{G}_M$ is an order filter in $\mathfrak{F}_M$;
2. if $A \notin \mathfrak{G}_M$ and $A \cup a, A \cup b \in \mathfrak{G}_M$ for some elements $a, b \in S$, $|B| = |A| + 1$ and $B \subseteq A \cup a \cup b$, then $B \in \mathfrak{G}_M$.

The following statement is valid.

Statement 16[11]. *A mapping* $\varphi_W : M \to H$ *is weak if and only if* $\mathfrak{F}_M - \mathfrak{F}_H$ *is a weak cut of matroid* M.

As any strong mapping is a weak mapping if there is a weak mapping between the matroids M and H, we shall use the notation $M \xrightarrow{\mathfrak{G}} H$ and say that it is generated by a weak cut $\mathfrak{G}_M$.

Statement 17. *If* $M \xrightarrow{\Phi} H$ *is an elementary mapping generated by a modular filter* Φ_M, *then the corresponding weak cut would be of the type* $\Phi_M \cap \mathfrak{F}_M$.

Proof. Suppose a subset $A \in \Phi_M \cap \mathfrak{F}_M$. As from (2) it follows that $\Phi_M = \{B \subseteq S \,|r_H(B) < r_M(B)\}$, so that $r_H(A) < |A| = r_M(A)$, and therefore, $A \notin \mathfrak{F}_H$. Hence $A \in \mathfrak{F}_M - \mathfrak{F}_H$ and $\Phi_M \cap \mathfrak{F}_M \subseteq \mathfrak{F}_M - \mathfrak{F}_H$.

Conversely, suppose a subset $A \in \mathfrak{F}_M - \mathfrak{F}_H$. Then $r_H(A) < |A|$ and $r_M(A) = |A|$, i.e., $r_H(A) < r_M(A)$ and therefore, $A \in \Phi_M$ and $A \in \Phi_M \cap \mathfrak{F}_M$.□

Theorem 1. *If* $M \xrightarrow{\mathfrak{G}} H$ *is a weak mapping generated by a weak cut* $\mathfrak{G}_M$, *then the following equation is valid:*

$$\mathfrak{G}_M = J(\mathfrak{B}_M - \mathfrak{B}_H) \cap \Phi(\mathfrak{R}_H - \mathfrak{R}_M)$$

Proof. By definition 23 and statement 16, $\mathfrak{G}_M = \mathfrak{F}_M - \mathfrak{F}_H = J(\mathfrak{B}_M) - J(\mathfrak{B}_H)$.

Suppose $A \in \mathfrak{G}_M$. As the subset $A \notin J(\mathfrak{B}_H)$ so $A \in \Phi(\mathfrak{R}_H)$, because it is $\mathfrak{R}_H$-dependent in the matroid H. Likewise, from $A \in J(\mathfrak{B}_M)$ it follows that $A \notin \Phi(\mathfrak{R}_M)$. Therefore, $A \in J(\mathfrak{B}_M) \cap \Phi(\mathfrak{R}_H)$ and $A \notin J(\mathfrak{B}_H) \cap \Phi(\mathfrak{R}_M)$. As under a weak mapping the inclusions $J(\mathfrak{B}_H) \subseteq J(\mathfrak{B}_M)$ and $\Phi(\mathfrak{R}_M) \subseteq \Phi(\mathfrak{R}_H)$ are valid, the subset A meets the condition $A \in \mathfrak{P} = J(\mathfrak{B}_M - \mathfrak{B}_M \cap \mathfrak{B}_H) \cap \Phi(\mathfrak{R}_H - \mathfrak{R}_H \cap \mathfrak{R}_M)$.

Let us show that for any weak mappings the equation $= J(\mathfrak{B}_M - \mathfrak{B}_H) \cap \Phi(\mathfrak{R}_H - \mathfrak{R}_M)$ is valid.

Suppose the ranks of matroids M and H are equal. Then $\mathfrak{B}_M \cap \mathfrak{B}_H = \mathfrak{B}_H$ and $J(\mathfrak{B}_M - \mathfrak{B}_M \cap \mathfrak{B}_H) = J(\mathfrak{B}_M - \mathfrak{B}_H)$. Then, if the mapping $M \xrightarrow{\mathfrak{G}} H$ is strong, $\mathfrak{B}_M \cap \mathfrak{B}_H = \varnothing$ and

$J(\mathfrak{B}_M - \mathfrak{B}_M \cap \mathfrak{B}_H) = J(\mathfrak{B}_M) = J(\mathfrak{B}_M - \mathfrak{B}_H)$. Similarly, the equation $\Phi(\mathfrak{R}_H - \mathfrak{R}_H \cap \mathfrak{R}_M) = \Phi(\mathfrak{R}_H - \mathfrak{R}_M)$ is fulfilled, regardless of whether $\mathfrak{R}_H \cap \mathfrak{R}_M \neq \varnothing$ or $\mathfrak{R}_H \cap \mathfrak{R}_M = \varnothing$.

Thus, an inclusion $\mathfrak{G}_M \subseteq J(\mathfrak{B}_M - \mathfrak{B}_H) \cap \Phi(\mathfrak{R}_H - \mathfrak{R}_M)$ is valid, from which, as the reverse $J(\mathfrak{B}_M - \mathfrak{B}_H) \cap \Phi(\mathfrak{R}_H - \mathfrak{R}_M) \subseteq \mathfrak{G}_M$ is obvious, follows the proof of the initial statement. □

3.2. The Weak Order on the Set of Matroids $\mathfrak{M}(S)$

Matroids of the same rank as opposed to strong mappings can be connected by weak mappings. Therefore, the set-theoretical inclusion of families of independent sets $\mathfrak{F}_M$ of matroids $M \in \mathfrak{M}(S)$ generates partial order, which is called *the order of a weak mapping* (further the *weak order)* on the whole set of matroids $\mathfrak{M}(S)$. The maximal element in a weak order is the free matroid $B(S)$, and the minimal is the matroid $B^*(S)$ of rank zero, in which all the elements of the set S are loops and there are no independent subsets.

If a matroid M covers a matroid H in a week order on the set $\mathfrak{M}(S)$, then the weak mapping $\varphi_W : M \to H$ is called a *simple weak mapping* (further a *simple mapping).*

Suppose the weak mapping $M \to M_1$ is generated by the weak cut $\mathfrak{G}_1 = \mathfrak{F}_M - \mathfrak{F}_{M_1}$, and the weak mapping $M \to M_2$ is generated by the weak cut $\mathfrak{G}_2 = \mathfrak{F}_M - \mathfrak{F}_{M_2}$. If $\mathfrak{F}_{M_2} \subset \mathfrak{F}_{M_1}$, then, by definition, there exists a weak mapping $M_1 \to M_2$, generated by the weak cut $\mathfrak{G}_3 = \mathfrak{F}_{M_1} - \mathfrak{F}_{M_2} = \mathfrak{G}_2 - \mathfrak{G}_1$. Hence, $\mathfrak{G}_1 \subset \mathfrak{G}_2$ and $\mathfrak{G}_3 \subset \mathfrak{G}_2$.

On the other hand, if the cuts $\mathfrak{G}_1$ and $\mathfrak{G}_2$ correspond to simple mappings, then $\mathfrak{G}_1 \not\subset \mathfrak{G}_2$, and $\mathfrak{G}_2 \not\subset \mathfrak{G}_1$, and matroids M_1 and M_2 are non-comparable in weak order on the set of matroids $\mathfrak{M}(S)$. Thus, a weak order on the set of matroids $\mathfrak{M}(S)$ can be defined as a set-theoretical inclusion of families of independent sets $\mathfrak{F}_M$, as well as an aggregate of all the weak cuts of the free matroid $B(S)$, ordered by inclusion.

Suppose now that $M \xrightarrow{\Phi_1} M_1$ and $M \xrightarrow{\Phi_2} M_2$ are elementary mappings, generated by modular filters Φ_1 and Φ_2 of the matroid M. From statement 17 it follows that the weak cuts $\mathfrak{G}_1 = \Phi_1 \cap \mathfrak{F}_M$ and $\mathfrak{G}_2 = \Phi_2 \cap \mathfrak{F}_M$ correspond to these elementary mappings as weak mappings. Therefore, if $\Phi_1 \subset \Phi_2$, then $\mathfrak{G}_1 \subset \mathfrak{G}_2$, and there exists a weak mapping $\varphi_W : M_1 \to M_2$ generated by the weak cut $\mathfrak{G}_3 = \mathfrak{G}_2 - \mathfrak{G}_1 = (\Phi_2 - \Phi_1) \cap \mathfrak{F}_M$. So, the set of all the modular filters of the matroid M, ordered by inclusion, generates a weak order on the set of all the corresponding elementary factors of the matroid M. Moreover, it can be shown that this partial order is also a lattice [2]. On the contrary, an aggregation of elementary factors of the matroid M is partially ordered, as are all the matroids from the set $\mathfrak{M}(S)$, by a weak order, and, therefore, the corresponding lattice is isomorphic to the lattice of modular filters of matroid M, ordered by inclusion.

A slightly different situation occurs on the set of all the non-trivial erections of an erectable matroid M. The rank of the erections is the same, obviously, and equals $r_M(S)+1$. Suppose a matroid H is a non-trivial erection of the matroid M. The matroid M is the top reduction of the matroid H and is generated by a modular cut $\mathfrak{Y}_H = S$, so that the respective modular filter Φ_H coincides with the set of generating subsets of the *set* S in H. As has already been pointed out, in this case the minimal elements of the order filter Φ_H are bases in the family $\mathfrak{B}_H$, i.e., $\Phi_H = \Phi(\mathfrak{B}_H)$. Thus, taking statement 17 in account, we obtain that in the case of the top reduction $\mathfrak{G} = \Phi(\mathfrak{B}_H) \cap \mathfrak{F}_H = \mathfrak{B}_H$.

Now suppose $M_1 \xrightarrow{\mathfrak{G}_1} M$ and $M_2 \xrightarrow{\mathfrak{G}_2} M$ are the top reductions of matroid M_1 and M_2, or (which is the same thing) suppose that matroids M_1 and M_2 are two different and non-trivial erections of the matroid M. From the a bove-said, $\mathfrak{G}_1 = \mathfrak{B}_{M_1}$ and $\mathfrak{G}_2 = \mathfrak{B}_{M_2}$. If $\mathfrak{B}_{M_2} \subset \mathfrak{B}_{M_1}$, then, similarly to the previous, there exists a weak mapping $M_1 \xrightarrow{\mathfrak{G}_3} M_2$, generated by the weak cut $\mathfrak{G}_3 = \mathfrak{B}_{M_1} - \mathfrak{B}_{M_2}$. Therefore, a set-theoretical inclusion of a family of bases generates a weak order on the set of all the non-trivial erections of any erectable matroid $M \in \mathfrak{M}(S)$. As with the lattice of modular filters, ordered by inclusion, it can be said that a set of all the non-trivial erections, ordered by inclusion, would be a lattice [2].

3.3. The Free Erection of Matroids

Suppose now that a matroid H is a non-trivial erection of a matroid M and that the top reduction mapping $H \xrightarrow{\mathfrak{G}} M$ in a weak order is simple. As $\mathfrak{G} = \mathfrak{B}_H$, the family of bases of matroid H cannot have a subfamily of bases of any matroid of the same rank, that contains the bases $\mathfrak{B}_M$ as subsets. In the contrary case there exists a weak image of the matroid H, the top reduction of which is the matroid M, and thus, the mapping $H \xrightarrow{\mathfrak{G}} M$ is not simple in weak order.

In other words, a so-called *free erection* is matched by the matroid H, or a point in the lattice of all erections of the matroid M, ordered by inclusion, the zero element of which is the trivial erection, which is the matroid M itself. This means that the matroid H as a free erection of the matroid M is covered by some matroid of the same rank as H in weak order of all erections of the matroid M. If we now assume that the matroid H is erectable, it also has its free erection, which must cover it in weak order, where this matroid has higher rank than the matroid H. From this fact, Kennedy [8] obtained the following result.

Statement 18. *A top reduction of any erectable matroid cannot be* a *simple mapping.*

In turn, Lucas [9] has shown that under a rank-preserving weak mapping $\varphi_W : M \to H$ uniforms matroids as minors of matroids M and H coincide. As uniforms matroids $M(n,k)$ for $n \geq 4$ and $2 \leq k \leq n-2$ are non-binary, hence this is a very important result for further study.

Statement 19. *The image of a binary matroid under a rank-preserving mapping is a binary matroid.*

It follows from these two statements that weak order in the category of matroids and their mappings is characterized by the property that the mappings of their top reduction correspond with their covers that lower the rank of the initial matroid, and that these covers alternate with the rank-preserving covers.

At this, if the initial matroid is binary, then the rank-preserving covers also preserve the binarity (the property of the matroid being binary), at the same time as the mappings of the top reduction not only change its rank, but can also lead to the loss of its binarity.

These results show that the study of the properties of matroids that are connected by morphisms in the category of matroids and their mappings, and, therefore, comparable in weak order on the set $\mathfrak{M}(S)$, comes down to the study of the properties of some subsets or the cuts of their families of cycles. This particular fact allows us to introduce the notion of “pseudo-matroids”, generated by elementary mappings of matroids, the families of cycles of which coincide with the minimal by inclusion elements of the abovementioned cuts.

***

In conclusion, let us highlight the thematic nature of this work once again. The information presented in the first part is mainly for reference and plays an introductory part for the following description and study of pseudo-matroids, generated by elementary mappings of matroids.

Chapter 2

PSEUDO-MATROIDS AND SEMI-MATROIDS

In the first part we have presented the known results of the theory of matroids that are mainly concerned with mappings as elements of the respective category. Further, we shall revisit these questions many times, but from a different perspective, based on the analysis of the properties of minimal by inclusion elements of modular filters, which define the properties of the family of cycles of pseudo-matroids generated by elementary mappings of matroids.

Part II consists of two sections dedicated to pseudo-matroids, generated by elementary mappings of matroids, and their special case of semi-matroids, generated by some G-mappings of binary matroids, respectively.

1. PSEUDO-MATROIDS GENERATED BY ELEMENTARY MAPPINGS OF MATROIDS

1.1. Weak Cuts and Modular Filters

According to statement 16, weak mappings of matroids are fully described by the weak cuts of the families of their independent subsets. Therefore, as strong mappings are also weak mappings, elementary mappings are defined not only by modular filters, but also by some weak cuts. Further, we shall look into the set-theoretical interconnections of such weak cuts and modular filters.

Every modular filter of any matroid from the set $\mathfrak{M}(S)$ as an order filter in the lattice $\mathbb{B}(S)$ is defined by its minimal elements of the structure, which is described in the theorem below.

Theorem 2. *If* $M \xrightarrow{\Phi} H$ *is an elementary mapping generated by a modular filter* Φ_M, *then the equation*

$\Phi_M = \Phi(\mathfrak{R}_H - \mathfrak{R}_M)$ is valid.

Proof. By the theorem's statement, $\mathfrak{B}_M \cap \mathfrak{B}_H = \varnothing$ and therefore $J(\mathfrak{B}_M - \mathfrak{B}_H) = J(\mathfrak{B}_M) = \mathfrak{F}_M$, so from theorem 1 it follows that the respective weak cut is of the type $\mathfrak{F}_M \cap \Phi(\mathfrak{R}_H - \mathfrak{R}_M)$. Hence, and from statement 17, we obtain that $\Phi_M = \Phi(\mathfrak{R}_H - \mathfrak{R}_M)$.□

It follows from theorems 1 and 2, then, that for an elementary mapping $M \xrightarrow{\Phi} H$ the cycles of the set $\mathfrak{R}_H - \mathfrak{R}_M$ would be minimal by inclusion elements of the modular filter Φ_M, as well as of the weak cut $\mathfrak{F}_M - \mathfrak{F}_H$.

Let us note that for any matroid M and cycles $C_1, C_2 \in \mathfrak{R}_M$, from the property of semimodularity of the rank function for matroids and from $|C_1 \cup C_2| = |C_1| + |C_2| - |C_1 \cap C_2|$, we obtain the inequation:

$$r_M(C_1 \cup C_2) \le |C_1 \cup C_2| - 2 \tag{7}$$

Further on, the proof of the known result [12] is carried out, based on theorem 2.

Theorem 3. *If* $M \xrightarrow{\Phi} H$ *is an elementary mapping generated by a modular filter* Φ_M *and* $\mathfrak{R}_M \cap \mathfrak{R}_H \neq \mathfrak{R}_M$ *then for any cycle* $C \in \mathfrak{R}_M - \mathfrak{R}_H$ *there exist cycles* $D_1, D_2 \in \mathfrak{R}_H - \mathfrak{R}_M$, *such that* $C = D_1 \cup D_2$.

Proof. Suppose $\mathfrak{R}_M \cap \mathfrak{R}_H \neq \mathfrak{R}_M$ and the cycle $C \in \mathfrak{R}_M - \mathfrak{R}_H$. As $C \in \mathfrak{R}_M$, then by definition $a \in \mathfrak{I}_M(C-a)$, for any element $a \in C$. As under a strong mapping $\mathfrak{I}_M(C-a) \subseteq \mathfrak{I}_H(C-a)$, then $a \in D \subseteq (C-a) \cup a = C$. Obviously, in this case certainly $D \in \mathfrak{R}_H - \mathfrak{R}_M$ and, therefore, $D \subset C$. Such a cycle exists for any element $a \in C$, so $C = \bigcup D$ is a union of not less than two cycles from the family $\mathfrak{R}_H - \mathfrak{R}_M$. Hence, $C \in \Phi(\mathfrak{R}_H - \mathfrak{R}_M) = \Phi_M$ and, therefore, $r_H(C) = r_M(C) - 1 = |C| - 2$. Now, if we suppose that $C = \bigcup D$ is a union of three or more cycles from the family $\mathfrak{R}_H - \mathfrak{R}_M$, taking (7) into account, we will obtain that $r_H(C) < |C| - 2$.□

From theorem 3, in particular, it follows that the flat $\mathfrak{I}_H(C)$ would be cyclic in matroid H for any cycle $C \in \mathfrak{R}_M - \mathfrak{R}_H$.

Theorem 4. *If $M \xrightarrow{\mathfrak{Y}} H$ is an elementary mapping generated by a modular cut $\mathfrak{Y}_M$, then $\mathfrak{R}_M \cap \mathfrak{R}_H \neq \varnothing$ if and only if in the family of flats $\mathfrak{I}_M$ there are cyclic flats that do not belong to the modular cut $\mathfrak{Y}_M$.*

Proof. Suppose that $\mathfrak{R}_M \cap \mathfrak{R}_H \neq \varnothing$. Then any cycle $C \in \mathfrak{R}_M \cap \mathfrak{R}_H$ does not belong to the modular filter Φ_M, which corresponds with the modular cut that generates the elementary mapping $M \xrightarrow{\mathfrak{Y}} H$. As for any matroid the closure of any cycle is a cyclic flat, the necessity is proven.

Conversely, suppose the family $\mathfrak{I}_M$ contains a cyclic flat that does not belong to the cut $\mathfrak{Y}_M$. As a cyclic flat is a union of cycles, there would be a cycle $C \in \mathfrak{R}_M$ that does not belong to the modular filter Φ_M. According to the property 2) of statement 15, there is always a cycle $D \in \mathfrak{R}_H$, such that $D \subseteq C$. Suppose $D \subset C$. As $D \notin \Phi_M$, $r_M(D) = r_H(D)$, must be true. However, $r_M(D) = |D| > r_H(D) = |D| - 1$. Therefore, $D = C$ and $C \in \mathfrak{R}_M \cap \mathfrak{R}_H$.□

Theorem 5. *A set S would be the only cyclic flat of a matroid M if and only if $|C| = r_M(S) + 1$ for all the cycles $C \in \mathfrak{R}_M$.*

Proof. Suppose that an elementary mapping $M \xrightarrow{\mathfrak{Y}} H$ is a top reduction mapping, so that $\mathfrak{Y}_M = S$ and $\Phi_M = \Phi(\mathfrak{B}_M)$. If $|C| = r_M(S) + 1$, where the cycle $C \in \mathfrak{R}_M$, then it is evident that $C \in \mathfrak{R}_M - \mathfrak{R}_H$ and, hence $C \notin \mathfrak{R}_H$. As $\Phi_M = \Phi(\mathfrak{R}_H - \mathfrak{R}_M)$, from here we get that for the top reduction $\mathfrak{R}_H = \mathfrak{B}_M \cup \{C \in \mathfrak{R}_M \mid |C| \leq r_M(S)\}$ and $\mathfrak{R}_M \cap \mathfrak{R}_H \neq \varnothing$, if there are cycles $C \in \mathfrak{R}_M$ such that $|C| \leq r_M(S)$. Thus, according to theorem 4, all cyclic flats of matroid M would belong to the modular cut $\mathfrak{Y}_M = S$ if and only if $|C| = r_M(S) + 1$ for all cycles $C \in \mathfrak{R}_M$.□

It follows from the proof of theorem 5 that, if there are no cycles $C \in \mathfrak{R}_M$, such that $|C| = r_M(S) + 1$ in matroid M, then for the top reduction it is true not only that $\mathfrak{R}_H - \mathfrak{R}_M = \mathfrak{B}_M$, but also that $\mathfrak{R}_M = \mathfrak{R}_H - \mathfrak{B}_M$.

1.2. Pseudo-Matroids and Their Bases

For any elementary factor H of the matroid M let us introduce the notation $\mathbb{R}_M(H) = \mathfrak{R}_H - \mathfrak{R}_M$.

According to theorem 2, the modular filter of any elementary mapping $M \xrightarrow{\Phi} H$ is of the form $\Phi_M = \Phi(\mathfrak{R}_H - \mathfrak{R}_M)$, and, therefore, is generated by cycles from $\mathbb{R}_M(H)$. If $\mathfrak{R}_H \cap \mathfrak{R}_M = \varnothing$, then $\Phi_M = \Phi(\mathfrak{R}_H)$ and the minimal by inclusion elements of the modular filter Φ_M coincide with the set $\mathfrak{R}_H$, and the modular filter Φ_M itself coincides with the family of all $\mathfrak{R}_H$-dependent sets of the matroid H. If $\mathfrak{R}_H \cap \mathfrak{R}_M \neq \varnothing$, then the set $\mathbb{R}_M(H)$ will not be a family of cycles of some matroid from the set $\mathfrak{M}(S)$. Further, we will show that both of these situations can be united, provided by introducing more general algebraic objects - pseudo-matroids generated by elementary mappings of matroids.

Suppose that $\mathfrak{A}$ is an arbitrary set of subsets of the set S, such that $A \not\subseteq B$ for all the subsets $A, B \in \mathfrak{A}$. For any subset $A \subseteq S$ let us define the closure operation $\mu_{\mathfrak{A}} : 2^S \to 2^S$ as follows:

$$\mu_{\mathfrak{A}}(A) = \{a \in S \,|\, a \in A \text{ or } a \in C \subseteq A \cup a \,,\ C \in \mathfrak{A}\}. \tag{8}$$

Let us note that if $\mathfrak{A} = \mathfrak{R}_M$ is a family of cycles of the matroid $M \in \mathfrak{M}(S)$, then $\mu_{\mathfrak{A}}(A) = \mathfrak{I}_M(A), A \subseteq S$.

Let us now show that the closure operation (8) fulfills the exchange property. Suppose elements $a, b \in S$, such that a $a \notin \mu_{\mathfrak{A}}(A)$ and $a \in \mu_{\mathfrak{A}}(A \cup b)$. Then, according to (8), there would be a subset $C \in \mathfrak{A}$, for which $a \in C \subseteq (A \cup b) \cup a$, and $C \not\subseteq A \cup a$ at that. Thus, $a, b \in C$ and $b \in C \subseteq (A \cup a) \cup b$ and hence $b \in \mu_{\mathfrak{A}}(A \cup a)$.

Later, we will show that if the closure operation (8) fulfills the *idempotency property,* i.e., $\mu_{\mathfrak{A}}(A) = \mu_{\mathfrak{A}}(\mu_{\mathfrak{A}}(A))$ for all subsets $A \subseteq S$, it would be a closure operator for some matroid from the set $\mathfrak{M}(S)$.

Before describing the conditions of idempotency for the closure operation (8), let us show that there is an equivalent form for the axiom of cycles of matroids.

Statement 20. *The conditions below are equivalent for a family of cycles* $\mathfrak{R}_M$ *of a matroid* $M \in \mathfrak{M}(S)$ *:*

1. $C_1, C_2 \in \mathfrak{R}_M, a \in C_1 \cap C_2 \Rightarrow \exists C \in \mathfrak{R}_M$ *and* $C \subseteq C_1 \cup C_2 - a$ *;*
2. $C_1, C_2 \in \mathfrak{R}_M, a \in C_1 \cap C_2$ *and* $b \in C_1 \oplus C_2 \Rightarrow \exists C \in \mathfrak{R}_M$ *and* $b \in C \subseteq C_1 \cup C_2 - a.$

Proof. Obviously, 1) follows from 2). Let us show the converse. Suppose cycles $C_1, C_2 \in \mathfrak{R}_M$ and elements $a \in C_1 \cap C_2$ and $b \in C_1 \oplus C_2$ are set. Suppose that all the cycles $C \in \mathfrak{R}_M$, such that $C \subseteq C_1 \cup C_2 - a$ do not contain the element b, i.e., $C \subseteq C_1 \cup C_2 - a - b$ and the cycles C_1 and C_2 have a minimal cardinality of $|C_1 \cup C_2|$ under this condition. With no loss of generality, suppose that $b \in C_1 - C_2$. Taking into account that $C_1 \not\subset C$ by the definition of cycles, and $C_1 \neq C$, as $b \notin C$, there would be an element $d \in C - C_1$. Because $d \in C_2 - C_1$, so $d \in C_2 \cap C, a \in C_2 - C$ and $|C_2 \cup C| < |C_1 \cup C_2|$. Therefore, according to the assumption of the minimality of $|C_1 \cup C_2|$, there exists a cycle $D \in \mathfrak{R}_M$, such that $a \in D \subseteq C_2 \cup C - d$. However, there cannot be a cycle that is contained in the set $D \cup C_1 - a$, which contains the element $b \in C_1 - D$ at the same time, as then it would also exist for the cycles C_1 and C_2. On the other hand, $|D \cup C_1| < |C_1 \cup C_2|$, and we arrive at a contradiction with the minimality of $|C_1 \cup C_2|$. □

Let us go back to the closure operation (8) and its idempotency conditions.

Statement 21. A *closure operation* $A \to \mu_{\mathfrak{A}}(A)$, $A \subseteq S$, *of the form* $\mu_{\mathfrak{A}}(A) = \{a \in S \,|\, a \in A$ or $a \in C \subseteq A \cup a\}$ fulfills *the idempotency property if and only if the family* $\mathfrak{A}$ *is a family of cycles of some matroid* $\mathfrak{M}(S)$.

Proof. Suppose that the family of subsets $\mathfrak{A}$ is a family of cycles of a matroid from the set $\mathfrak{M}(S)$ and $\mu_{\mathfrak{A}}(A) \neq \mu_{\mathfrak{A}}(\mu_{\mathfrak{A}}(A))$ for some subset $A \subseteq S$. Because for any element $a \in A$ it is true that $a \in \mu_{\mathfrak{A}}(A)$ and $a \in \mu_{\mathfrak{A}}(\mu_{\mathfrak{A}}(A))$, the inequation $\mu_{\mathfrak{A}}(A) \neq \mu_{\mathfrak{A}}(\mu_{\mathfrak{A}}(A))$ means that there exist elements $a,b \in S$, such that $a,b \notin A, a \in \mu_{\mathfrak{A}}(A)$ and $b \notin \mu_{\mathfrak{A}}(A)$, and $b \in \mu_{\mathfrak{A}}(\mu_{\mathfrak{A}}(A) \cup a)$ at that. In other words, there would be found subsets $C_1, C_2 \in \mathfrak{A}$, for which $a \in C_1 \subseteq A \cup a$, $a,b \in C_2 \subseteq A \cup a \cup b$ and $C_2 \not\subseteq A \cup b$.

Obviously, $a \in C_1 \cap C_2$ and $b \in C_2 - C_1$. Therefore, by the condition 2) of statement 20, there would be a subset $C \in \mathfrak{A}$, such that $b \in C \subseteq C_1 \cup C_2 - a \subseteq A \cup b$. Thus, $b \in \mu_{\mathfrak{A}}(A)$, which contradicts the assumption.

Conversely, suppose that for some subsets $C_1, C_2 \in \mathfrak{A}$ and an element $a \in C_1 \cap C_2$ the subset $C_1 \cup C_2 - a$ does not contain any subsets from the set $\mathfrak{A}$. Then for any element $b \in C_1 \oplus C_2$ the condition $b \notin \mu_{\mathfrak{A}}(C_1 \cup C_2 - a - b)$ holds true. With no loss of generality, let us assume that $b \in C_2 - C_1$. Then $b \in \mu_{\mathfrak{A}}(C_2 - b)$, because $b \in C_2 \subseteq (C_2 - b) \cup b = C_2$. Therefore, $b \in \mu_{\mathfrak{A}}(C_1 \cup C_2 - b)$. Analogously, $a \in \mu_{\mathfrak{A}}(C_1 - a)$ and $a \in \mu_{\mathfrak{A}}(C_1 \cup C_2 - a - b)$. The latter means that $C_1 \cup C_2 - b \subseteq \mu_{\mathfrak{A}}(C_1 \cup C_2 - a - b)$. So, we arrive at $b \in \mu_{\mathfrak{A}}(C_1 \cup C_2 - b) \subseteq \mu_{\mathfrak{A}}(\mu_{\mathfrak{A}}(C_1 \cup C_2 - a - b))$, and the closure operation (8) for the set $C_1 \cup C_2 - a - b \subseteq S$ is not idempotent. □

Suppose that $M \xrightarrow{\Phi} H$ is an elementary mapping, generated by a modular filter $\Phi_M = \Phi(\mathbb{R}_M(H))$. If $\mathfrak{R}_H \cap \mathfrak{R}_M = \varnothing$, then $\mathbb{R}_M(H) = \mathfrak{R}_H$, and from statement 21 it follows that the closure operation (8) fulfills the idempotency property. If $\mathfrak{R}_H \cap \mathfrak{R}_M \neq \varnothing$, then $\mathbb{R}_M(H)$ would not be a family of cycles of a matroid from the set $\mathfrak{M}(S)$.

Definition 24. *A pseudo-matroid* $\mathbb{R}(M,H)$ *is an ordered couple* $< S, \mathbb{R}_M(H) >$, *where* $\mathbb{R}_M(H)$ *is a family of cycles of a pseudo-matroid.*

As with matroids, any subset $A \subseteq S$ we shall call $\mathbb{R}_M(H)$*-independent* in the pseudo-matroid $\mathbb{R}(M,H)$ if it does not contain cycles from $\mathbb{R}_M(H)$ as subsets. In the contrary case, this set would be $\mathbb{R}_M(H)$-dependent. Respectively, the maximal by inclusion $\mathbb{R}_M(H)$-independent subsets $B \subseteq S$, such that $\mu_{\mathbb{R}_M(H)}(B) = S$, would be the bases of the pseudo-matroid $\mathbb{R}(M,H)$. We shall denote the family of bases of a pseudo-matroid by $\mathbb{B}_M(H)$. In particular, if $\mathfrak{R}_H \cap \mathfrak{R}_M = \varnothing$, then $\mathbb{R}_M(H) = \mathfrak{R}_H$ and $\mathbb{B}_M(H) = \mathfrak{B}_H$. In the general case the following result is valid.

Theorem 6. If $\mathfrak{K}_M$ and $\mathfrak{K}_H$ are the *families of co-points of matroids* $M, H \in \mathfrak{M}(S)$, which are *connected by an elementary mapping* $M \xrightarrow{\Phi} H$, *then the set* $\mathbb{B}_M(H) = \mathfrak{K}_M - \mathfrak{K}_H$ *is a family of bases of the pseudo-matroid* $\mathbb{R}(M,H)$.

Proof. Suppose that a subset $B \subseteq S$ is a base of the pseudo-matroid $\mathbb{R}(M,H)$. According to $\mathbb{R}_M(H) = \mathfrak{R}_H - \mathfrak{R}_M$ we get that $\mathfrak{I}_H(B) = S$ and $r_H(B) = r_H(S) = r_M(S) - 1$. Hence, $B \not\subseteq K$, where $K \in \mathfrak{K}_H$, and, as the mapping is elementary, there would be found a co-point $K \in \mathfrak{K}_M - \mathfrak{K}_H$, such that $B \subseteq K$. As the co-point $K \notin \Phi_M = \Phi(\mathbb{R}_M(H))$, it is a $\mathbb{R}_M(H)$-independent set and, considering the maximality by inclusion, we arrive at $B = K$ and $\mathbb{B}_M(H) \subseteq \mathfrak{K}_M - \mathfrak{K}_H$.

Conversely, for any co-point $K \in \mathfrak{K}_M$ the equations $r_M(K) = r_M(S) - 1 = r_H(S)$ are valid. If $K \in \mathfrak{K}_M - \mathfrak{K}_H$ then $r_H(K) = r_M(K) = r_H(S)$ and, hence, the co-point K is a generating subset of the set S in matroid H. At that $K \notin \Phi_M = \Phi(\mathbb{R}_M(H))$, i.e., the set $K \subseteq S$ would be $\mathbb{R}_M(H)$-independent.

If $a \in S - K$, then $r_M(K \cup a) = r_M(S) > r_H(K \cup a) = r_H(S)$. Therefore, $K \cup a \in \Phi(\mathbb{R}_M(H))$, i.e., it is a $\mathbb{R}_M(H)$-dependent set. From this it follows that $\mathfrak{K}_M - \mathfrak{K}_H \subseteq \mathbb{B}_M(H)$ and, therefore, we arrive at the fact that any co-point $K \in \mathfrak{K}_M - \mathfrak{K}_H$ is a base of the pseudo-matroid $\mathbb{R}(M,H)$ and there are no other bases in the family $\mathbb{B}_M(H)$.□

For any elementary mapping $M \xrightarrow{\mathfrak{Y}} H$, generated by a modular cut $\mathfrak{Y}_M$, the co-points that belong to it form a linear cut $\mathfrak{L}_M$ of the family of co-points $\mathfrak{K}_M$. It thus follows from theorem 6 that the co-points from the subfamily $\mathfrak{K}_M - \mathfrak{L}_M$ would be the bases $\mathbb{B}_M(H)$ of the respective pseudo-matroid, and that $\mathfrak{L}_M = \mathfrak{K}_M \cap \mathfrak{K}_H$.

As a corollary theorem 6, considering the note made above, we arrive at the following result.

Theorem 7. *For an arbitrary matroid* $M \in \mathfrak{M}(S)$ *an elementary factor* $H \in \mathfrak{M}(S)$, *which fulfills the condition* $\mathfrak{R}_H \cap \mathfrak{R}_M = \varnothing$, *exists if and only if* $\mathfrak{K}_M - \mathfrak{K}_H = \mathfrak{B}_H$.

Let us note that, as with matroids, theorem 6 gives a description of the families of $\mathbb{R}_M(H)$-independent and $\mathbb{R}_M(H)$-dependent sets of the pseudo-matroid $\mathbb{R}(M,H)$ as elements of the ideal, generated by bases $\mathbb{B}_M(H)$, and of the filter, generated by cycles $\mathbb{R}_M(H)$, respectively. At the same time, let us highlight that, in the case of $\mathfrak{R}_M \cap \mathfrak{R}_H \neq \varnothing$, it follows from statement 21 that in the pseudo-matroid $\mathbb{R}(M,H)$ there would always be a subset $A \subseteq S$, such that $\mu_{\mathbb{R}_M(H)}(A) \neq \mu_{\mathbb{R}_M(H)}(\mu_{\mathbb{R}_M(H)}(A))$. At that, for any base $B \in \mathbb{B}_M(H)$ of the pseudo-matroid, the idempotency property is fulfilled, because from the equation $\mu_{\mathbb{R}_M(H)}(B) = S$ it inevitably follows that $\mu_{\mathbb{R}_M(H)}(B) = \mu_{\mathbb{R}_M(H)}(\mu_{\mathbb{R}_M(H)}(B))$. Thus, as opposed to matroids, in the general case, such notions as the rank and closure of an arbitrary subset $A \subseteq S$ are not defined unequivocally for pseudo-matroids $\mathbb{R}(M,H)$.

1.3. Pseudo-Matroids and the Factorization of Strong Mappings

In the theory of matroids the systems of axioms that define, for example, families of bases, cycles or co-points, do not include an algorithmic component that would allow for the construction of such families. In other words, they only provide the possibility of checking whether a given set of subsets complies with the axioms of matroids. In such a situation only a procedure based on the enumeration of systems of subsets of the set S and then checking whether every such option fulfills certain properties can be suggested for the construction of matroids, and this is hardly very different from full enumeration. In this connection, the development of algorithms for the construction of matroids that are fundamentally different from enumeration and have a substantially lower computational complexity presents a great interest. From here on we shall refer to such algorithms as "constructive".

If any given matroid H is viewed as an elementary factor of some matroid M, then to construct M would mean to add to the family of flats $\Im_H$ the subfamilies $\Im_M - \Im_H$ of subsets of the set S, such that the set $\Im_M$ would be a family of flats of matroid M of the rank $r_H(S)+1$. Let us remind ourselves that such matroid M would be a lift of matroid H in matroid M.

From theorem 6 we get that, under any elementary mapping, the bases $\mathbb{B}_M(H)$ of the corresponding pseudo-matroid $\mathbb{R}(M,H)$ are the maximal by inclusion elements of the subfamily of flats $\Im_M - \Im_H$. Considering this, and also knowing the form of generating elements of any modular filter $\Phi_M = \Phi(\mathbb{R}_M(H))$, we can look upon the procedure of construction of lifts M of a given matroid H, described above, from a different, non-classical, point of view.

Indeed, we can build a family of cycles $\mathfrak{R}_M$ of some matroid M through a family of cycles $\mathfrak{R}_H$, so that the filter $\Phi_M = \Phi(\mathbb{R}_M(H))$ would be a modular filter, and therefore, $M \xrightarrow{\Phi} H$ would be an elementary mapping. Let us show that, based on this procedure, it is possible to construct a family of cycles of a Higgs lift in an explicit form under a canonical mapping of $B(S) \to H$.

Theorem 8. *Under a canonical mapping of* $B(S) \to H$ a *matroid* M *would be a Higgs lift of a matroid* H *of the rank* $r_H(S) \le |S|-2$ *if and only if the family of its cycles is defined by the equation* $\mathfrak{R}_M = \{C = D_1 \cup D_2 \mid D_1, D_2 \in \mathfrak{R}_H \text{ and } C - \min\}$

Proof. Suppose that under a canonical mapping of $B(S) \to H$ the matroid M is a Higgs lift of the matroid H, so that $B(S) \to M \xrightarrow{\Phi} H$. If $C \in \mathfrak{R}_M \cap \mathfrak{R}_H$ then $C \notin \Phi_M = \Phi(\mathbb{R}_M(H))$, and therefore, $r_M(C) = r_H(C) = |C|-1$, which contradicts the condition 3) of statement 14, according to which any subset $C \subseteq S$ in Higgs factorization of a canonical mapping belongs to the modular filter Φ_M if $|C| > r_H(C)$. Therefore, $\mathfrak{R}_M \cap \mathfrak{R}_H = \varnothing$ and, from theorem 3 we get that for any cycle

$C \in \mathfrak{R}_M = \mathfrak{R}_M - \mathfrak{R}_H$ there would be cycles $D_1, D_2 \in \mathfrak{R}_H = \mathfrak{R}_H - \mathfrak{R}_M$, such that $C = D_1 \cup D_2$.

Conversely, for the family $\mathfrak{A} = \{C = D_1 \cup D_2 \mid D_1, D_2 \in \mathfrak{R}_H$ and $C - \min\}$ let us introduce the closure operation (8): for any subset $A \subseteq S$, $\mu_{\mathfrak{A}}(A) = \{a \in S \mid a \in A$ or $a \in C \subseteq A \cup a,\ C \in \mathfrak{A}\}$.

As $C = D_1 \cup D_2$ and $D_1, D_2 \in \mathfrak{R}_H$, $\mu_{\mathfrak{A}}(A) \subseteq \mathfrak{I}_H(A)$ for all subsets $A \subseteq S$. If the set $A \in \mathfrak{F}_H$, then, by the definition of the family $\mathfrak{A}$ and the axiom of cycles for matroids, we arrive at the equation $\mu_{\mathfrak{A}}(A) = A$ and, as a consequence, $\mu_{\mathfrak{A}}(A) = \mu_{\mathfrak{A}}(\mu_{\mathfrak{A}}(A))$. If $A \notin \mathfrak{F}_H$, then there exists a cycle $D_1 \in \mathfrak{R}_H$, such that $D_1 \subseteq A$. Then, for any element $a \notin A$ and $a \in \mathfrak{I}_H(A)$ from the condition $a \in D_2 \subseteq A \cup a$ it follows that $a \in C = D_1 \cup D_2 \subseteq A \cup a$, i.e., $a \in \mu_{\mathfrak{A}}(A)$. From this we infer that $\mathfrak{I}_H(A) \subseteq \mu_{\mathfrak{A}}(A)$. Thus, considering the above for the sets $A \notin \mathfrak{F}_H$, the equation $\mu_{\mathfrak{A}}(A) = \mathfrak{I}_H(A)$ is valid, and, as above, $\mu_{\mathfrak{A}}(A) = \mu_{\mathfrak{A}}(\mu_{\mathfrak{A}}(A))$. So, the closure operation (8) is idempotent; the family $\mathfrak{A}$, according to statement 21, is a family of cycles of some matroid $M \in \mathfrak{M}(S)$ and $\mu_{\mathfrak{A}}(A) = \mathfrak{I}_M(A), A \subseteq S$.

Let us note that, according to the definition of the family $\mathfrak{A}$, the equation $\mathfrak{R}_M \cap \mathfrak{R}_H = \varnothing$ is valid. As for any subset $A \subseteq S$, $\mathfrak{I}_M(A) \subseteq \mathfrak{I}_H(A)$, it follows from the condition 2) of statement 10 that there exists a strong mapping $\varphi_S : M \to H$. Any cycle $D \in \mathfrak{R}_H = \mathfrak{R}_H - \mathfrak{R}_M$ in the matroid M is a $\mathfrak{R}_M$-independent set, and, therefore, there would be a base $B \in \mathfrak{B}_M$ for which $D \subseteq B$. At that, for any base $B \in \mathfrak{B}_M$ there are no cycles $D_1, D_2 \in \mathfrak{R}_H$, such that $D_1 \cup D_2 \subseteq B$, because otherwise there would be a cycle $C \in \mathfrak{A}$, such that $C \subseteq B$ and $C \in \mathfrak{R}_M$ which is impossible. Therefore, the matroid M and H are connected not only by a strong mapping, but by an elementary strong mapping.

From what has been said earlier, $\mathfrak{I}_M(A) = \mathfrak{I}_H(A)$ if $A \notin \mathfrak{F}_H$, and $\mathfrak{I}_M(A) = A$, if $A \in \mathfrak{F}_H$. Therefore, $r_{B(S)}(\mathfrak{I}_M(A)) = r_H(\mathfrak{I}_M(A))$, for any flat $\mathfrak{I}_M(A) \in \mathfrak{I}_M - \mathfrak{I}_H$. Thus, by definition, the matroid M is not only a lift of the matroid H under a canonical mapping $B(S) \to H$, but it is a Higgs lift.□

There are two important inferences from theorem 8.

1. If under a canonical mapping of $B(S) \to H$ a matroid M is a Higgs lift of a matroid H, then $\mathfrak{R}_M \cap \mathfrak{R}_H = \varnothing$. Respectively, the lifts of the matroid H, for which $\mathfrak{R}_M \cap \mathfrak{R}_H \neq \varnothing$, are not Higgs lifts.
2. If a matroid H is binary, then under a canonical mapping its Higgs lift would not necessarily be a binary matroid.

The binarity of matroids is also not necessarily preserved under a top reduction mapping. The following theorem shows how the Higgs lifts and the mappings of the top reduction can be connected.

Theorem 9. *If* $M \xrightarrow{\Phi} H$ *is an elementary mapping, then* $\mathfrak{R}_M \cap \mathfrak{R}_H = \varnothing$ *if and only if an elementary mapping of the dual matroids* $H^* \xrightarrow{\Phi^*} M^*$ *is a top reduction mapping and* $\Phi^*_{H^*} = \Phi(\mathfrak{B}_{H^*})$.

Proof. Suppose that $\mathfrak{R}_M \cap \mathfrak{R}_H = \varnothing$. According to theorem 6, $\mathfrak{K}_M - \mathfrak{K}_H = \mathfrak{B}_H$ in the statement. From this and from the property 3) of statement 2 we obtain that $S - C \in \mathfrak{B}_H$ for any cycle $C \in \mathfrak{R}_{M^*} - \mathfrak{R}_{H^*}$. This means that $C \in \mathfrak{B}_{H^*}$ and the elementary mapping $H^* \xrightarrow{\Phi^*} M^*$ is generated, according to theorem 2, by the modular filter $\Phi^*_{H^*} = \Phi(\mathfrak{R}_{M^*} - \mathfrak{R}_{H^*}) = \Phi(\mathfrak{B}_{H^*})$. Therefore, the matroid M^* is the top reduction of the matroid H^*.

Conversely, suppose that the matroid M^* is the top reduction of the matroid H^*. Then $\mathfrak{K}_{M^*} \cap \mathfrak{K}_{H^*} = \varnothing$. As above, from here we obtain that for any cycles $C \in \mathfrak{R}_M$ and $D \in \mathfrak{R}_H$ the equation $S - C \neq S - D$ is valid, and therefore, so is the equation $C \neq D$. The latter means that $\mathfrak{R}_M \cap \mathfrak{R}_H = \varnothing$.□

From theorem 9 we obtain the criterion of erectability for matroids from the set $\mathfrak{M}(S)$.

Theorem 10. *An arbitrary matroid* $M \in \mathfrak{M}(S)$ *is erectable if and only if there exists an elementary factor* H^* *of the dual matroid* M^*, *such that* $\mathfrak{R}_{H^*} \cap \mathfrak{R}_{M^*} = \varnothing$.

Proof. Any elementary *factor* H^* of the matroid M^* generates a mapping $M^* \xrightarrow{\Phi^*} H^*$. If $\mathfrak{R}_{H^*} \cap \mathfrak{R}_{M^*} = \varnothing$, then, according to theorem 9, the respective mapping $H \xrightarrow{\Phi} M$ would be a top reduction mapping and, therefore, the matroid M would be erectable.

If the matroid M is erectable, then there would be found a matroid H, such that the mapping $H \to M$ would be a top reduction mapping. Hence, and from the proof of theorem 9, $\mathfrak{R}_{H^*} \cap \mathfrak{R}_{M^*} = \varnothing$.□

If the matroid H is the top reduction of the matroid M, then from theorem 5, $\mathfrak{R}_M \cap \mathfrak{R}_H = \varnothing$ if and only if $|C| = r_M(S) + 1$ for all cycles $C \in \mathfrak{R}_M$. Hence, the following result.

Theorem 11. *Elementary mappings* $M \xrightarrow{\Phi} H$ *and* $H^* \xrightarrow{\Phi^*} M^*$ *are the mappings of the top reduction if and only if* $|C| = r_M(S) + 1$ *and* $|D^*| = r_{H^*}(S) + 1$ *for all cycles* $C \in \mathfrak{R}_M$ *and* $D^* \in \mathfrak{R}_{H^*}$.

Theorem 8 implies that if the matroid M is a Higgs lift of the matroid H under a canonical mapping of $B(S) \to H$, then $\mathfrak{R}_M \cap \mathfrak{R}_H = \varnothing$ Therefore, *from* theorem 10 we obtain that, if the matroid M is *a* Higgs lift of some matroid H under a canonical mapping $B(S) \to H$, then the dual matroid M^* is *erectable.*

Let us illustrate the outcomes through an example.

Example 3. Suppose $S=\{1,2,3,4,5\}$. Consider a matroid $M \in \mathfrak{M}(S)$, defined as an ordered couple $< S, \mathfrak{R}_M >$ with a family of cycles $\mathfrak{R}_M = \{\{1,2,3\},\{3,4,5\},\{1,2,4,5\}\}$. The lattice of flats $\mathfrak{I}_M$ of the matroid M is presented in Picture 1.

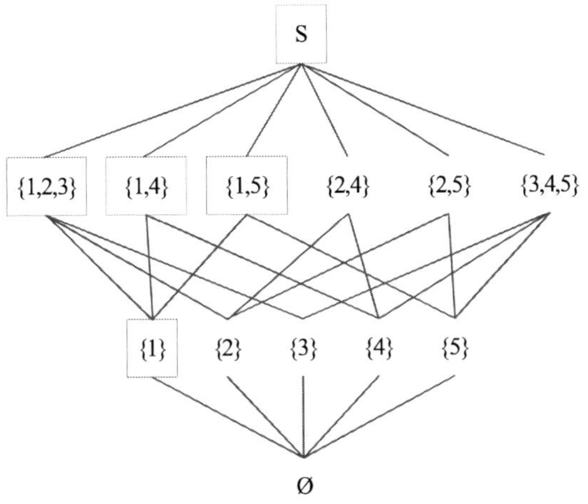

Picture 1. The lattice of flats $\mathfrak{I}_M$.

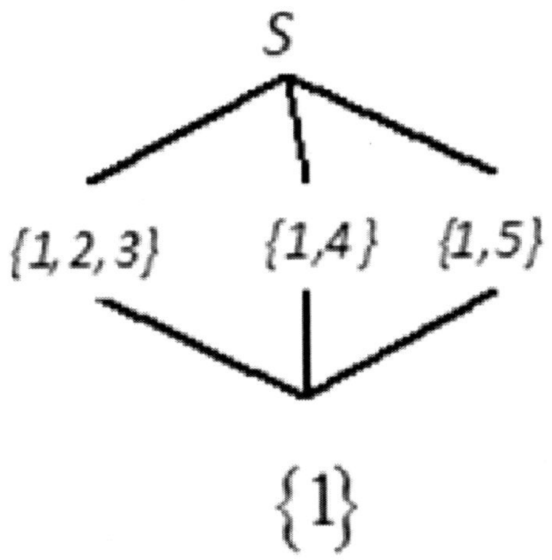

Picture 2. The lattice of flats $\mathfrak{I}_H$.

The set $\mathfrak{Y}_M = \{\{1\},\{1,2,3\},\{1,4\},\{1,5\},S\}$, according to definition 14, would be a modular cut of the family of flats $\mathfrak{I}_M$. The respective modular filter $\Phi_M = \Phi(\{1\},\{2,3\},\{2,4,5\})$. The modular cut $\mathfrak{Y}_M$ generates an elementary mapping of $M \xrightarrow{\mathfrak{Y}} H$, and the lattice of flats $\mathfrak{I}_H$ of an elementary factor H of the matroid M is presented in Picture 2.

It is easy to see that the set $\mathfrak{R}_H = \{\{1\},\{2,3\},\{2,4,5\},\{3,4,5\}\}$ would be a family of cycles of the matroid H. Respectively, $\mathbb{R}_M(H) = \mathfrak{R}_H - \mathfrak{R}_M = \{\{1\},\{2,3\},\{2,4,5\}\}$. Thus, the modular filter $\Phi_M = \Phi(\mathbb{R}_M(H))$ is generated by the cycles of the pseudo-matroid $\mathbb{R}(M,H)$.

From the type of the set $\mathbb{R}_M(H)$ it follows that the set $\mathbb{B}_M(H) = \{\{2,4\},\{2,5\},\{3,4,5\}\}$ would be the family of bases of pseudo-matroid $\mathbb{R}(M,H)$. According to Pictures 1 and 2, we get the equation $\mathbb{B}_M(H) = \mathfrak{K}_M - \mathfrak{K}_H$, which is in accordance with the theorem 6. Let us note that the sets $\{3, 4\}, \{3, 5\}$ and $\{4, 5\}$ in the pseudo-matroid $\mathbb{R}(M,H)$ are not idempotent. For example, $\mu_{\mathbb{R}_M(H)}(\{3,4\}) = \{1,3,4,5\}$, while $\mu_{\mathbb{R}_M(H)}(\mu_{\mathbb{R}_M(H)}(\{3,4\})) = S$.

Further, $\mathfrak{R}_M \cap \mathfrak{R}_H = \{3, 4, 5\}$ and, as $\mathfrak{R}_M \cap \mathfrak{R}_H \neq \varnothing$ then, according to theorem 4, there must exist a cyclic flat that does not belong to the modular cut $\mathfrak{Y}_M$. Indeed, the flat $\{3,4,5\}$ is cyclic in the matroid M and $\{3,4,5\} \notin \mathfrak{Y}_M$. In turn, the set $\mathfrak{R}_M - \mathfrak{R}_H = \{1,2,4,5\}$ and the cycle $\{1,2,4,5\} = \{1\} + \{2,4,5\}$, where cycles $\{1\},\{2,4,5\} \in \mathfrak{R}_H - \mathfrak{R}_M$ which is also in accordance with theorem *3*.□

To conclude, let us highlight that in the first section of this part we have introduced a new notion in the theory of matroids - pseudo-matroids generated by elementary mappings of matroids. It is important to note that these pseudo-matroids not only are “cryptomorphic” versions [3, 23] of matroids, defined by means of a closure operation $\mu_{\mathfrak{A}}$ for some family of subsets $\mathfrak{A} \subseteq 2^S$, but also are generated specifically by elementary mappings of matroids, and are therefore functionally Connected to the category of matroids and their mappings. Families of cycles of pseudo-matroids are minimal by inclusion elements of the respective modular filters, and so a pseudo-matroid unequivocally defines an elementary mapping by which it is generated. The investigation of the properties of pseudo-matroids brings about some new outcomes of a general-theoretical nature. In particular, the family of cycles of Higgs lifts of a matroid under a canonical mapping is explicitly described and the erectability criterion for matroids is obtained.

2. Semi-Matroids, Generated by *G*-Mappings of Binary Matroids

In this section a new class of elementary mappings in the theory of matroids that preserve binarity is introduced, which are called "*G*-mappings". The properties of semi-matroids, generated by *G*-mappings, are studied as a special case of pseudo-matroids. It is proven that, as opposed to pseudo-matroids, semi-matroids can be considered axiomatically defined algebraic objects, generated by the category of binary matroids and their mappings, and the category of binary matroids itself is closed under morphisms, *G*-mappings and rank-preserving weak mappings.

2.1. *G*-Mappings of Binary Matroids. Modular Filters of *G*-Mappings

According to the outcomes of the previous section, for arbitrary matroids $M \in \mathfrak{M}(S)$ the minimal by inclusion elements of weak cuts and modular filters, which generate elementary mappings $M \xrightarrow{\Phi} H$, are described through the families of cycles of the elementary factors H of the matroid M. Further, we will show that for binary matroids there exist elementary mappings for which analogous elements can be constructed directly from the family of independent sets $\mathfrak{F}_M$ of the matroid M.

Let us denote a family of binary matroids, defined on the set S, as $\mathfrak{N}(S) \subseteq \mathfrak{M}(S)$. From here on, *unless* specifically stated, we will assume that all the matroids considered belong to the set $\mathfrak{N}(S)$.

Let us define the families of subsets $\mathbb{R}_M(D_0)$ and $\mathfrak{R}_M(D_0)$ for a binary matroid M and a set $D_0 \in \mathfrak{F}_M$ as follows:

$$\mathbb{R}_M(D_0) = \{D_0\} + \{D \in \mathfrak{F}_M \mid D \oplus D_0 = \sum C, C \in \mathfrak{R}_M\} \tag{9}$$

and

$$\mathfrak{R}_M(D_0) = \{C \in \mathfrak{R}_M \mid D \not\subset C, D \in \mathbb{R}_M(D_0)\} \tag{10}$$

Let us note that if $D_0 = \varnothing$, then $\mathbb{R}_M(\varnothing) = \varnothing$ and $\mathfrak{R}_M(\varnothing) = \mathfrak{R}_M$.

It follows from (9) that the family $\mathbb{R}_M(D_0), D_0 \neq \varnothing,$ consists of all minimal by inclusion sets $D \in \mathfrak{F}_M$, such that $D = D_0 \oplus \sum C,$ where $D_0 \cap C \neq \varnothing$ and $C \in \mathfrak{R}_M$. It is obvious from the condition $D \in \mathfrak{F}_M$ that $D_0 \cap C \neq \varnothing$. Further, if there exists a subset $D' \subset D$, for which $D' = D_0 \oplus \sum C',$ where $D_0 \cap C' \neq \varnothing$ and $C' \in \mathfrak{R}_M$, then, according to the axiom of cycles for binary matroids $D \oplus D' = \sum C''$ and $C'' \in \mathfrak{R}_M$, which is impossible, as $D \oplus D' = D - D' \in \mathfrak{F}_M$.

From the above it follows that the family of subsets $\mathbb{R}_M(D_0)$ can be constructed by iterations. On the first step of iteration a family of subsets of the set S of the type $D = D_0 \oplus C$ is constructed, where the cycles $C \in \mathfrak{R}_M$ and $D_0 \cap C \neq \varnothing$. Of the following step, subsets of the type $D \oplus C$ are constructed in an analogous way, and so on. As the set $\mathfrak{R}_M$ is finite, and on every step of the iteration there would be only the minimal by inclusion sets left, there would definitely come a moment when the newly constructed family of subsets would coincide with the family obtained on the previous step.

Let us examine the second step of the iteration, and suppose the set $D = (D_0 \oplus C_1) \oplus C_2$, where $D_0 \cap C_1 \neq \varnothing$ and $(D_0 \oplus C_1) \cap C_2 \neq \varnothing$.

If $C_1 \cap C_2 = \varnothing$, then $D = D_0 \oplus (C_1 + C_2)$, and hence $D \in \mathbb{R}_M(D_0)$.

If $C_1 \cap C_2 \neq \varnothing$, then, according to the axiom of cycles for binary matroids, $C_1 \oplus C_2 = \sum C$. At that, according to the property 2) of statement 20, the latter sum would always contain a family of cycles that, in turn, contain any element $a \in C_1 - C_2$. In other words, there always exists a sum of mutually non-intersecting cycles $\sum C' \subseteq \sum C$, such that $D_0 \cap C' \neq \varnothing$, i.e., there exists a set $D' = D_0 \oplus \sum C' \in \mathbb{R}_M(D_0)$, for which the inclusion $D' \subseteq D$, is valid. And this means that, leaving only the minimal by inclusion sets on every step of the iteration, in the resulting population of subsets we will only get elements from the family $\mathbb{R}_M(D_0)$.

An analogous situation occurs on any of the following steps of the iteration. It therefore follows from what has just been said that the family of sets, constructed on the last step of the iteration, is exactly the set of subsets $\mathbb{R}_M(D_0)$.

Several properties of the family $\mathbb{R}_M(D_0)$ are joined together by the following theorem.

Theorem 12. *For any binary matroid* $M \in \mathfrak{N}(S)$ *and subset* $D_0 \in \mathfrak{F}_M, D_0 \neq \varnothing$, *the family of subsets* $\mathbb{R}_M(D_0)$ *fulfills the following properties:*

1. $D \in \mathbb{R}_M(D_0) \Rightarrow \mathbb{R}_M(D) = \mathbb{R}_M(D_0)$;
2. *for any* $B \in \mathfrak{B}_M$ *there exists a single* $D \subseteq B$, *such that* $D \in \mathbb{R}_M(D_0)$;
3. *for any* $B_1, B_2 \in \mathfrak{B}_M$ *there exists a single couple* $D_1 \subseteq B_1$ *and* $D_2 \subseteq B_2$, *such that* $\mathbb{R}_M(D_1) = \mathbb{R}_M(D_2)$;
4. $D_1, D_2 \subseteq B$, $B \in \mathfrak{B}_M$ *and* $D_1 \neq D_2 \Rightarrow \mathbb{R}_M(D_1) \neq \mathbb{R}_M(D_2)$.

Proof.

1. Suppose that $D \in \mathbb{R}_M(D_0)$. Obviously, $D \in \mathbb{R}_M(D)$. According to (9), the set $D_0 = D \oplus \sum C$, where $C \in \mathfrak{R}_M$ and $D \cap C \neq \varnothing$, as, by definition, $D_0 \in \mathfrak{F}_M$. As there are no subsets $D_0' \subset D_0$, such that $D_0' = D \oplus \sum C'$, where $C' \cap D \neq \varnothing$ and $C' \in \mathfrak{R}_M$, then $D_0 \in \mathbb{R}_M(D)$, and therefore, $\mathbb{R}_M(D_0) = \mathbb{R}_M(D)$.

2. Suppose $B \in \mathfrak{B}_M$. Let us assume that $D_0 \not\subseteq B$. Then, $D_0 - B \neq \varnothing$ and for any element $d \in D_0 - B$ there will always be found a single fundamental cycle $C(d,B) \in \mathfrak{R}_M$, such that $d \in C(d,B) \subseteq B \cup d$. Therefore, $D_0 \oplus \sum_{d \in D_0 - B} \oplus C(d,B) \subseteq B.$ From the axiom of cycles for binary matroids we get that $\sum_{d \in D_0 - B} \oplus C(d,B) = \sum C'$, and by construction in the direct sum $\sum C'$ there would always be contained a set of $|D_0 - B|$ of cycles C'_d, that contain all elements $d \in D_0 - B$. In other words, there would always exist a set $D = D_0 \oplus \sum_{d \in D_0 - B} C'_d \in \mathbb{R}_M(D_0)$, for which the inclusion $D \subseteq B$ is valid.

   Now, suppose there exist different sets $D_1, D_2 \in \mathbb{R}_M(D_0)$, such that $D_1, D_2 \subseteq B$ from some base $B \in \mathfrak{B}_M$. Then, by the definition of the family $\mathbb{R}_M(D_0)$, we get $D_1 \oplus D_2 = \sum C$, where $C \in \mathfrak{R}_M$ and $D_1 \oplus D_2 \subseteq B$, which is impossible.
3. From the property 2) it follows that, for a given set $D_0 \in \mathfrak{F}_M$, any base $B \in \mathfrak{B}_M$ contains a single subset $D \subseteq B$, such that $D \in \mathbb{R}_M(D_0)$. Thus, the proof of the property 3) follows directly from the property 1); in particular, from the validity of the equation $\mathbb{R}_M(D) = \mathbb{R}_M(D_0)$ for any subsets $D \in \mathbb{R}_M(D_0)$
4. Suppose that for some base $B_0 \in \mathfrak{B}_M$ the set $D_0 \subseteq B_0$. Let us choose a subset $D \subseteq B_0$ and $D \neq D_0$. From the above-proven property 2) it follows that $D \notin \mathbb{R}_M(D_0)$, and from property 1) – that $\mathbb{R}_M(D) \neq \mathbb{R}_M(D_0)$. Thus, the proof of the property 4) follows directly from the fact that D_0 can be any subset of the family $\mathfrak{F}_M$, and, therefore, can belong to any base from the set $\mathfrak{B}_M$. □

Theorem 13. *For any binary matroid* $M \in \mathfrak{N}(S)$ *and a subset* $D_0 \in \mathfrak{F}_M, D_0 \neq \varnothing$, *the family of the subsets* $\mathfrak{R} \subset 2^S$ *that fulfill the equations* $\mathfrak{R} - \mathfrak{R}_M = \mathbb{R}_M(D_0)$ *and* $\mathfrak{R} \cap \mathfrak{R}_M = \mathfrak{R}_M(D_0)$ *is a family of cycles of a binary matroid* $H \in \mathfrak{N}(S)$, *which is connected to matroid* M *by an elementary mapping* $M \xrightarrow{\Phi} H$, *generated by a modular filter* $\Phi_M = \Phi(\mathbb{R}_M(D_0))$.

Proof. Let us show that the family $\mathfrak{R}$, defined in the statement of theorem 13, is a family of cycles of some binary matroid $H \in \mathfrak{N}(S)$. Suppose $D_1, D_2 \in \mathfrak{R}$. If $D_1, D_2 \in \mathbb{R}_M(D_0) = \mathfrak{R} - \mathfrak{R}_M$, then, according to (9), $D_1 \oplus D_2 = \sum C$, where either $C \in \mathfrak{R}_M \cap \mathfrak{R} = \mathfrak{R}_M(D_0)$, or $C \in \mathfrak{R}_M - \mathfrak{R}_M(D_0)$. In the latter case there exists a subset $D_1^{'} \in \mathbb{R}_M(D_0)$, such that $D_1^{'} \subset C$. If we denote $D_2^{'} = C \oplus D_1^{'} = C - D_1^{'}$, the inclusion $D_2^{'} \in \mathbb{R}_M(D_0)$ will take place, and $C = D_1^{'} + D_2^{'}$. So, the binary sum $D_1 \oplus D_2$ is decomposed into the sum of the non-intersecting subset from the family $\mathfrak{R}$.

A similar outcome for the subsets $D_1, D_2 \in \mathfrak{R} \cap \mathfrak{R}_M = \mathfrak{R}_M(D_0)$ follows from the axiom of cycles for a binary matroid M as from the form of the cycles from the set $\mathfrak{R}_M - \mathfrak{R}_M(D_0)$.

If $D_1 \in \mathbb{R}_M(D_0)$ and $D_2 \in \mathfrak{R}_M(D_0)$, then, as $D_2 \in \mathfrak{R}_M$ and $D_2 \notin \mathfrak{R}_M - \mathfrak{R}$, either $D_1 \cap D_2 = \varnothing$ or $D_1 \cap D_2 \neq \varnothing$ and $D_1 \cap D_2 \neq D_1$. In the first case, $D_1 \oplus D_2 = D_1 + D_2$, where $D_1, D_2 \in \mathfrak{R}$, while in the second, by definition, $D_1 \oplus D_2 = D \in \mathbb{R}_M(D_1)$, and therefore $D \in \mathfrak{R}$. As a summary, the binary sum $D_1 \oplus D_2$ is decomposed into the sum of non-intersecting subsets from the family $\mathfrak{R}$. Thus, the family of subsets $\mathfrak{R}$ fulfill the axiom of cycles for some binary matroid $H \in \mathfrak{N}(S)$ and $\mathfrak{R} = \mathfrak{R}_H$.

Consider a set $A \subseteq S$ and a flat $\mathfrak{I}_M(A)$ in the matroid M. It follows from statement 21 that $\mathfrak{I}_M(A) = \{a \in S \,|\, a \in A$ or $a \in C \subseteq A \cup a$, $C \in \mathfrak{R}_M\}$.

If the cycle $C \in \mathfrak{R}_M(D_0) = \mathfrak{R}_H \cap \mathfrak{R}_M$, then, obviously, $C \in \mathfrak{R}_H$. If the cycle $C \in \mathfrak{R}_M - \mathfrak{R}_M(D_0)$, then $C = D_1 + D_2$, *where* $D_1, D_2 \in \mathbb{R}_M(D_0) = \mathfrak{R}_H - \mathfrak{R}_M$, i.e., the cycle C is again a direct sum of cycles $\mathfrak{R}_H$. In other words, in any case we get $\mathfrak{I}_M(A) \subseteq \mathfrak{I}_H(A)$. Considering the property 2) of statement 10, we infer that there exists a strong mapping $\varphi_S : M \to H$. The elementary character of this mapping follows, in turn, from the property 2) of theorem 12. Thus, the matroids M and H are connected by an elementary mapping $M \xrightarrow{\Phi} H$, which, according to theorem 2, is generated by the modular filter $\Phi_M = \Phi(\mathfrak{R}_H - \mathfrak{R}_M) = \Phi(\mathbb{R}_M(D_0))$.□

Definition 25. An elementary mapping $M \xrightarrow{\Phi} H$ of binary matroids $M, H \in \mathfrak{N}(S)$, generated by a modular filter $\Phi_M = \Phi(\mathbb{R}_M(D_0))$ for some subset $D_0 \in \mathfrak{F}_M$, $D_0 \neq \varnothing$, we shall call *a G-mapping,* and the matroid H - a *G-factor* of the matroid M.

A G-factor H of an arbitrary matroid $M \in \mathfrak{N}(S)$, generated by some subset $D_0 \in \mathfrak{F}_M, D_0 \neq \varnothing$, we shall further refer to as $H(D_0)$.

Let us note that all the loops of a binary matroid M belong to the family of cycles $\mathfrak{R}_M(D_0)$ for any set $D_0 \in \mathfrak{F}_M, D_0 \neq \varnothing$, by definition, and, therefore, also to any of its G-factors. Also, note that a G-factor of a binary matroid is its elementary factor.

By definition, the set $\{H(D_0) | D_0 \subseteq B, D_0 \neq \varnothing\}$ is a set of all G-factors of the matroid M for a given base $B \in \mathfrak{B}_M$ and, according to theorem 12, it would be the same for any bases $B \in \mathfrak{B}_M$. Thus, the family of all G-factors of an arbitrary matroid $M \in \mathfrak{N}(S)$, defined as an ordered couple $< S, \mathfrak{R}_M >$, is constructed by means of iteration in the way described above. At this, the subsets $D_0 \subseteq B, D_0 \neq \varnothing$, for any fixed $B \in \mathfrak{B}_M$ unequivocally define the factors $H(D_0)$ of the matroid M in the respective pseudo-matroids $\mathbb{R}(M, H(D_0))$.

Definition 26. For any binary matroids $M, H(D_0) \in \mathrm{N}(S)$, $D_0 \in \mathfrak{F}_M, D_0 \neq \varnothing$, connected by a G-mapping, the pseudo-matroid $\mathbb{R}(M, H(D_0))$ we shall call a *semi-matroid* and denote by $\mathbb{R}(M, D_0)$.

According to definition 25 and theorem 13, the family of subsets, defined by the equation (9) would be the family of cycles of the semi-matroid $\mathbb{R}(M, D_0)$. Respectively, the set $\mathbb{B}_M(D_0) = \mathfrak{K}_M - \mathfrak{K}_{H(D_0)}$ would be the family of bases of the semi-matroid $\mathbb{R}(M, D_0)$.

The characteristic property that differentiates a semi-matroid from a pseudo-matroid is that the binary sum of cycles from $\mathbb{R}_M(D_0)$ is not a cycle of the semi-matroid $\mathbb{R}(M, D_0)$. Further, we shall demonstrate the crucial importance of this property.

Theorem 14. *For the binary matroids* $M, H \in \mathfrak{N}(S)$ *matroid* H *is a G-factor of the matroid* M *if and only if* $D_1 \oplus D_2 = \sum C$, $C \in \mathfrak{R}_M$, *for* any *different cycles* $D_1, D_2 \in \mathfrak{R}_H - \mathfrak{R}_M$, *and for any cycle* $C \in \mathfrak{R}_M - \mathfrak{R}_H$ *the flat* $\mathfrak{I}_H(C)$ *is cyclic in matroid* H.

Proof. If the matroid H is a G-factor of the matroid M, then the proof of necessity follows from theorem 13. In particular, if $C \in \mathfrak{R}_M - \mathfrak{R}_H$, then the equation $C = D_1 + D_2$, where $D_1, D_2 \in \mathfrak{R}_H - \mathfrak{R}_M$, is valid, and, therefore, the flat $\mathfrak{I}_H(C)$ is cyclic in the matroid H. Let us now prove the sufficiency.

For any cycles $C \in \mathfrak{R}_M$ either $C \in \mathfrak{R}_M \cap \mathfrak{R}_H$ or $C \in \mathfrak{R}_M - \mathfrak{R}_H$. If $C \in \mathfrak{R}_M - \mathfrak{R}_H$, then by assumption the flat $\mathfrak{I}_H(C)$ is cyclic in the matroid H and, therefore, $C = \bigcup D$, where, obviously, the cycles $D \in \mathfrak{R}_H - \mathfrak{R}_M$.

Thus, the cycle C is a union of cycles from the set $\mathfrak{R}_H$. As in the proof of theorem 13, it follows from this that the matroids M and H are connected by a strong mapping, as for any subset $A \subseteq S$ the inclusion $\mathfrak{I}_M(A) \subseteq \mathfrak{I}_H(A)$ is valid.

Moreover, this mapping is elementary, because any base $B \in \mathfrak{B}_M$ cannot contain more than one cycle from the family $\mathfrak{R}_H - \mathfrak{R}_M$, as otherwise their binary sum contains a cycle of the matroid M, which is impossible.

By assumption, if the cycles $D_0, D \in \mathfrak{R}_H - \mathfrak{R}_M$, then the equation $D_0 \oplus D = \sum C$, $C \in \mathfrak{R}_M$ is valid. Then $D_0 \cap C \neq \varnothing$ and $D \cap C \neq \varnothing$ for all such cycles C, otherwise either $C \subseteq D_0$ or $C \subseteq D$, which is impossible, as $D, D_0 \in \mathfrak{F}_M - \mathfrak{F}_H$. Therefore, $D = D_0 \oplus \sum C \in \mathbb{R}_M(D_0)$ for any cycles $D \in \mathfrak{R}_H - \mathfrak{R}_M$. And as, by definition, $D_0 \in \mathbb{R}_M(D_0)$, then $\mathfrak{R}_H - \mathfrak{R}_M \subseteq \mathbb{R}_M(D_0)$ for any cycle $D_0 \in \mathfrak{R}_H - \mathfrak{R}_M$.

On the other hand, the matroid H is an elementary factor of the matroid M, and from theorem 3, $C = D_1 \cup D_2$ for any cycle $C \in \mathfrak{R}_M - \mathfrak{R}_H$, where cycles $D_1, D_2 \in \mathfrak{R}_H - \mathfrak{R}_M$. As $D_1 \oplus D_2 \subseteq C$ and, by assumption $D_1 \oplus D_2 = \sum C'$, $C' \in \mathfrak{R}_M$, the latter inclusion is only possible when $C = D_1 + D_2$. Therefore, for any cycle $C \in \mathfrak{R}_M - \mathfrak{R}_H$ there exists a cycle $D_1 \in \mathfrak{R}_H - \mathfrak{R}_M$, *such that* $C \cap D_1 \neq \varnothing$ and $C \oplus D_1 = D_2 \in \mathfrak{R}_H - \mathfrak{R}_M$.

Suppose now that the cycle $C \in \mathfrak{R}_H \cap \mathfrak{R}_M$ and $C \cap D \neq \varnothing$ for some $D \in \mathfrak{R}_H - \mathfrak{R}_M$. Then, by the axiom of cyc1es for binary matroids, $D \oplus C = \sum D'$, where for any cycle in the direct sum either $D' \in \mathfrak{R}_H - \mathfrak{R}_M$, or $D' \in \mathfrak{R}_H \cap \mathfrak{R}_M$. In the first case $D \oplus D' \subseteq C$ which, by assumption, is possible if and only if $D \oplus C = D' \in \mathfrak{R}_H - \mathfrak{R}_M$. If the cycle $D' \in \mathfrak{R}_H \cap \mathfrak{R}_M$, then, on the contrary $D' \oplus C \subseteq D$, and, by the axiom of cycles for binary matroids, $D' \oplus C = \sum C'$, $C' \in \mathfrak{R}_M$, which is impossible, as the inclusion $D \in \mathfrak{F}_M - \mathfrak{F}_H$ is valid.

Summarizing, we find that for any cycle $C \in \mathfrak{R}_M$, such that $C \cap D \neq \varnothing$, where $D \in \mathfrak{R}_H - \mathfrak{R}_M$, there exists a cycle $D' = C \oplus D$, and $D' \in \mathfrak{R}_H - \mathfrak{R}_M$. Therefore, as the set $D \in \mathbb{R}_M(D_0)$ is of the form $D = D_0 \oplus \sum_{i=1}^{k} C_i$, where $C_i \cap D_0 \neq \varnothing$, the equations $D = D_0 \oplus \sum_{i=1}^{k} C_i = D_1 \oplus \sum_{i=2}^{k} C_i = ... = D_{k-1} \oplus C_k = D_k \in \mathfrak{R}_H - \mathfrak{R}_M$ hold true.

Hence, $\mathbb{R}_M(D_0) \subseteq \mathfrak{R}_H - \mathfrak{R}_M$ and, considering what has just been said, $\mathfrak{R}_H - \mathfrak{R}_M = \mathbb{R}_M(D_0)$.

It has been stated above that any cycle $C \in \mathfrak{R}_M - \mathfrak{R}_H$ does not belong to the family $\mathfrak{R}_M(D_0)$, defined by the expression (10), as there would always be found a cycle $D \in \mathbb{R}_M(D_0)$, such that $D \subset C$. This fact concludes the proof of necessity.□

Theorem 14 holds true for any binary matroids $M, H \in \mathfrak{N}(S)$. However, if it is known that the matroids are connected by an elementary mapping, then the respective conditions can be simplified.

Theorem 15. *If the binary matroids $M, H \in \mathfrak{N}(S)$ are connected by an elementary mapping $M \xrightarrow{\Phi} H$, then the matroid H is a G-factor of the matroid M if and only if $D_1 \oplus D_2 = \sum C$, $C \in \mathfrak{R}_M$ for any different cycle $D_1, D_2 \in \mathfrak{R}_H - \mathfrak{R}_M$.*

Proof. As for any elementary mapping $M \xrightarrow{\Phi} H$ the respective modular filter is $\Phi_M = \Phi(\mathfrak{R}_H - \mathfrak{R}_M)$ and for any cycle $C \in \mathfrak{R}_M - \mathfrak{R}_H$, according to theorem 3, the equation $C = D_1 \cup D_2$, where cycles $D_1, D_2 \in \mathfrak{R}_H - \mathfrak{R}_M$, hold true, and the proof of the statement immediately follows from theorem 14.□

These outcomes allow the introduction of the notion of a semi-matroid, defined as an ordered couple $< S, \mathbb{R}_M(D_0) >$, as an axiomatically set algebraic object, generated by a binary matroid from the set $\mathfrak{N}(S)$

For the given $M \in \mathfrak{N}(S)$ let us fix the base $B \in \mathfrak{B}_M$, and suppose $\mathfrak{A}_M(B) = \{\mathbb{R}_M(D_0) | D_0 \subseteq B, D_0 \neq \varnothing\}$. From theorem 12 it follows that $\mathfrak{A}_M(B_1) = \mathfrak{A}_M(B_2)$ for any bases $B_1, B_2 \in \mathfrak{B}_M$, while from theorem 13 it follows that $\mathfrak{A}_M(B) \subseteq \{\mathfrak{R}_{H(D_0)} - \mathfrak{R}_M | D_0 \subseteq B, D_0 \neq \varnothing$ and $H(D_0)$ is a G*-factor* of $M\}$. On the other hand, for any semi-matroid $\mathbb{R}(M, D_0)$, considered as a pseudo-matroid $\mathbb{R}(M, H(D_0))$, the family of its cycles is

$\mathbb{R}_M(D_0) = \mathfrak{R}_{H(D_0)} - \mathfrak{R}_M \subseteq \mathfrak{F}_M - \mathfrak{F}_{H(D_0)}$ and, therefore, considering theorem 15, the inclusion $\mathbb{R}_M(D_0) \in \mathfrak{A}_M(B)$ is valid for some base $B \in \mathfrak{B}_M$, such that $D_0 \subseteq B$.

To summarize, every binary matroid $M \in \mathfrak{N}(S)$ for any base $B \in \mathfrak{B}_M$ generates a family of $2^{r_M(S)} - 1$ semi-matroids, defined as ordered couple $< S, \mathbb{R}_M(D_0) >$, where $D_0 \subseteq B, D_0 \neq \varnothing$ and $\mathbb{R}_M(D_0) \in \mathfrak{A}_M(B)$, and there are no more such semi-matroids. Therefore, the subset $\mathfrak{A}_M \subseteq \mathfrak{F}_M$ would be a family of cycles of a semi-matroid, generated by a matroid $M \in \mathfrak{N}(S)$ and defined as an ordered couple $< S, \mathfrak{A}_M >$ if and only if $\mathfrak{A}_M = \mathbb{R}_M(D_0)$ for some $D_0 \in \mathfrak{F}_M, D_0 \neq \varnothing$. Therefore, the condition $D \in \mathbb{R}_M(D_0), D_0 \in \mathfrak{F}_M, D_0 \neq \varnothing$, where $\mathbb{R}_M(D_0) = \{D_0\} + \{D \in \mathfrak{F}_M | D \oplus D_0 = \sum C, C \in \mathfrak{R}_M\}$ can be considered as the *axiom of cycles* for semi-matroid $\mathbb{R}(M, D_0)$, generated by binary matroids $M \in \mathfrak{N}(S)$.

Hence, the introduction of the notion of G-mappings of binary matroids allows us to construct a family of semi-matroids, defined on the set S, as an algebraic structure, generated axiomatically by the set of binary matroids $\mathfrak{N}(S)$.

2.2. G-Factorization of Binary Matroids

Let us now consider the issues related to the possibility of construction of all the binary matroids by means of consequential G-mappings of a free matroid $B(S)$.

Theorem 16. *All the matroids on the set* $\mathfrak{N}(S)$ *of rank* $|S|-1$ *are* G-*factors of a free matroid* $B(S)$.

Proof. A free matroid $B(S)$ is binary, as there are no cycles in it. Any matroid of rank $|S|-1$ is also binary, as its family of cycles consists of one subset $D \subseteq S$. The proof of the statement follows from the property 4) of theorem 12, i.e., from the fact that $\mathbb{R}_{B(S)}(D_1) \neq \mathbb{R}_{B(S)}(D_2)$ for any two different subsets $D_1, D_2 \subseteq S$. □

As a corollary of theorem 16, we find that in the family $\mathfrak{M}(S)$ of all matroids, defined on the set S, there exists precisely $2^{|S|}-1$ matroids of rank $|S|-1$, all of them binary.

For a binary matroid H let us consider factorization into elementary mappings of a canonical mapping $B(S)\to H$ of the form

$$B(S)=H_0\xrightarrow{\Phi_1}H_1\xrightarrow{\Phi_2}\ldots\xrightarrow{\Phi_{k-1}}H_{k-1}\xrightarrow{\Phi_k}H_k=H \quad (11)$$

Definition 27. As a G*-factorization* let us refer to the factorization (11) of a canonical mapping $B(S)\to H$, where matroids H_i are G-factors of matroids H_{i-1}, $i=\overline{1,k}$.

Let us highlight that, according to definition 20, the matroid H_{i-1} cannot be a lift of the matroid H_i under a canonical mapping of $B(S)\to H_i$, $i=\overline{1,k}$. Therefore, a question arises: does G-factorization of a canonical mapping $B(S)\to H$ exist for all binary matroids H?

In theorem 9 for arbitrary matroids $M,H\in\mathfrak{M}(S)$, connected by an elementary mapping, the case of $\mathfrak{R}_M\cap\mathfrak{R}_H=\varnothing$ has been considered. It was shown that in this case an elementary mapping of dual matroids is a top reduction mapping. It is known that a top reduction mapping may not preserve the matroid's binarity. Moreover, the contrary situation is possible, when the result of the top reduction of a non-binary matroid would be a binary matroid. This is particularly relevant for matroids of ranks 1 and $|S|-1$, which are binary, but can be elementary factors of some non-binary matroids or contain non-binary matroids as elementary factors. For example, if $|S|=n$ then the top reduction of a non-binary uniform matroid $M(n,2)$ is a binary matroid $M(n,1)$. At the same time a non-binary dual matroid $M^*(n,2)=M(n,n-2)$ is the top reduction of a binary matroid $M^*(n,1)=M(n,n-1)$. It is clear that $\mathfrak{R}_{M(n,2)}\cap\mathfrak{R}_{M(n,1)}=\mathfrak{R}_{M(n,n-2)}\cap\mathfrak{R}_{M(n,n-1)}=\varnothing$, and all the listed matroids fulfill the condition of theorem 10.

Let us denote $\mathfrak{N}_k(S)=\{M\in\mathfrak{N}(S)\,|\,r_M(S)=k\}$.

Theorem 17. *If* $M \in \mathfrak{N}_{|S|-1}(S)$, H *is a* G*-factor of matroid* M, $\mathfrak{R}_M = \{C\}$ *and* $\mathfrak{R}_H - \mathfrak{R}_M = \mathbb{R}_M(D_0)$ *for some subset* $D_0 \in \mathfrak{F}_M, D_0 \neq \varnothing$, *then* $\mathfrak{R}_M \cap \mathfrak{R}_H = \varnothing$ *if and only if* $D_0 \subset C$.

Proof. If $\mathfrak{R}_M = \{C\}$ and $D_0 \subset C$, then the definition of the families $\mathbb{R}_M(D_0)$ and $\mathfrak{R}_M(D_0)$ implies that $\mathbb{R}_M(D_0) = \{D_0, C - D_0\}$ and $\mathfrak{R}_M(D_0) = \varnothing$. As $\mathfrak{R}_M(D_0) = \mathfrak{R}_M \cap \mathfrak{R}_H$, according to theorem 13, $\mathfrak{R}_M \cap \mathfrak{R}_H = \varnothing$.

Conversely, suppose that $\mathfrak{R}_M \cap \mathfrak{R}_H = \mathfrak{R}_M(D_0) = \varnothing$ and $D_0 \not\subset C$. Then, regardless of whether $D_0 \cap C = \varnothing$ or $D_0 \cap C \neq \varnothing$, the cycle C, by definition, belongs to the family $\mathfrak{R}_M(D_0)$, which contradicts the assumption.□

Let us show, then, that in the condition of theorem 17 the respective top reduction of dual matroids would also be a G-mapping.

Theorem 18. *If a binary matroid* $H \in \mathfrak{N}(S)$ *is a* G*-factor of a matroid* $M \in \mathfrak{N}_{|S|-1}(S)$ *and* $\mathfrak{R}_M \cap \mathfrak{R}_H = \varnothing$, *then the matroid* M^* *would be a* G*-factor of the matroid* H^*.

Proof. Suppose that $M \in \mathfrak{N}(S)$, $r_M(S) = |S| - 1$, $r_H(S) = |S| - 2$ and $\mathfrak{R}_H - \mathfrak{R}_M = \{D_0, C - D_0\}$, where $\mathfrak{R}_M = \{C\}$, the subset $D_0 \in \mathfrak{F}_M, D_0 \neq \varnothing$, and $D_0 \subset C$. Then the dual matroid M^* of rank 1 has a family of cycles $\mathfrak{R}_{M^*} = \{\{a\} \mid a \in S - C\} + \{\{a, b\} \mid a, b \in C\}$, and, therefore, a family of bases $\mathfrak{B}_{M^*} = \{\{a\} \mid a \in C\}$.

As, according to theorem 9, dual matroids are connected by an elementary mapping $H^* \xrightarrow{\Phi^*} M^*$, which is a top reduction, then $\Phi^*_{H^*} = \Phi(\mathfrak{B}_{H^*})$ and in the conditions of theorem not only is $\mathfrak{R}_{M^*} - \mathfrak{R}_{H^*} = \mathfrak{B}_{H^*}$, but also $\mathfrak{R}_{H^*} = \mathfrak{R}_{M^*} - \mathfrak{B}_{H^*}$. Indeed, for any base $B^* \in \mathfrak{B}_{H^*}$, $|B^*| = 2$ the conditions $|B^* \cap D_0| = 1$ and $|B^* \cap (C - D_0)| = 1$ hold true, as in the contrary case either $B^* \subseteq D_0$ and, therefore, $C - D_0 \subseteq S - B^*$, or $B^* \subseteq C - D_0$ and $D_0 \subseteq S - B^*$, i.e., the set $B = S - B^*$ would not be a base of matroid H.

Therefore, there cannot be any cycles the length of $r_{H^*}(S)+1=3$ in matroid H^*.

As a result, we arrive at

$$\mathfrak{R}_{H^*}=\{\{a\}\mid a\in S-C\}+\{\{a,b\}\mid\{a,b\}\subseteq D_0 \text{ or } \{a,b\}\subseteq C-D_0\}.$$

Thus, for any bases $B_1^*,B_2^*\in\mathfrak{B}_{H^*}$ the equation $B_1^*\oplus B_2^*=\sum D^*$, $D^*\in\mathfrak{R}_{H^*}$ is valid, as, if $B_1^*=\{a_1,b_1\}$, $B_2^*=\{a_2,b_2\}$, where $a_1,a_2\in D_0$ and $b_1,b_2\in C-D_0$, then either $B_1^*\oplus B_2^*=\{a_1,a_2\}$ or $B_1^*\oplus B_2^*=\{b_1,b_2\}$ or $B_1^*\oplus B_2^*=\{a_1,a_2\}+\{b_1,b_2\}$. As $\mathfrak{R}_{M^*}-\mathfrak{R}_{H^*}=\mathfrak{B}_{H^*}$, the proof of this statement follows from theorem 15.□

Theorem 18 means that when a binary matroid H of rank $|S|-2$ is a G-factor of a binary matroid M of rank $|S|-1$ and $\mathfrak{R}_M\cap\mathfrak{R}_H=\varnothing$, the top reduction mapping $H^*\to M^*$ would also be a G-mapping.

Definition 28. Suppose $S=S_1+S_2+...+S_k$, where $|S_i|\geq 1$, $S_i\cap S_j=\varnothing$ for all $1\leq i<j\leq k$. A *matroid M* is *representable as a sum of matroids* $M=M_1+M_2+...+M_k$, where $M_i\in\mathfrak{M}(S_i)$, $i=\overline{1,k}$, if for any base $B\in\mathfrak{B}_M$ the equation $B=B_1+B_2+...+B_k$ is valid and $B_i\in\mathfrak{B}_{M_i}, i=\overline{1,k}$.

For example, if $S=\{a_1,a_2,...,a_n\}$, then a free matroid $B(S)$, according to definition 28, is *representable as a sum of matroids* $B(S)=M(|a_1|,1)+M(|a_2|,1)+...+M(|a_n|,1)$, where $M(|a_i|,1)$ is a uniform matroid of rank 1, defined on the single-element set of $a_i, i=\overline{1,n}$.

Theorem 19. *A top reduction mapping of a binary matroid* $M\in\mathfrak{N}(S)$ *without loops and of rank* $r_M(S)=k$ *would be a* G*-mapping if and only if* $S=S_1+S_2+...+S_k$, *where* $|S_i|\geq 1$, $S_i\cap S_j=\varnothing$ *for all* $1\leq i<j\leq k$, *and the equation holds true.*

$$M = M(|S_1|,1) + M(|S_2|,1) + \ldots + M(|S_k|,1) \qquad (12)$$

Proof. Suppose that a binary matroid M is connected. Suppose an elementary mapping $M \xrightarrow{\Phi} H$ is a top reduction mapping and a G-mapping. Then $\mathfrak{R}_H - \mathfrak{R}_M = \mathfrak{B}_M$ and, according to theorem 15, for any bases $B_1, B_2 \in \mathfrak{B}_M$ the condition $B_1 \oplus B_2 = \sum C$, $C \in \mathfrak{R}_M$ is fulfilled. For any base $B_1 \in \mathfrak{B}_M$ and an element $a_1 \in S - B_1$ there exists a single fundamental cycle $C(a_1, B_1)$, such that $a_1 \in C(a_1, B_1) \subseteq B_1 \cup a_1$. Then for any element $a_2 \in C(a_1, B_1) \cap B_1$ the set $B_2 = (B_1 - a_2) \cup a_1$ belongs to the family $\mathfrak{B}_M$, i.e., it would be a base of the matroid M. Therefore, $B_1 \oplus B_2 = \{a_1, a_2\}$ and $\{a_1, a_2\} \in \mathfrak{R}_M$. Thus, all fundamental cycles for any base of the matroid M are of the length 2. As the equation (1) holds true for binary matroids, all cycles from the family $\mathfrak{R}_M$ would be the same. As the matroid M is connected, the matroid $M = M(|S|,1)$ is uniform of rank 1, defined on the whole set S. Therefore, if the rank is $r_M(S) = k > 1$, then the matroid M is separable. By definition of connectedness, mutually non-intersection subsets of the set S correspond to connectivity components. As any set $S' \subset S$ would be a connectivity component if and only if the matroid $M|S'$ is connected, considering what has just been said, (12) is proven true.

Conversely, the family of flats of matroids $M(|S_i|,1)$ is of the form $\mathfrak{I}_{M(|S_i|,1)} = \{\varnothing, S_i\}$ for all $i = \overline{1,k}$. Therefore, the lattice $\mathfrak{I}_M$ of flats of the matroid M is isomorphic to the lattice $\mathbb{B}(\{S_1, S_2, \ldots, S_k\})$ of a free matroid of the rank k, its points being the sets $S_i, i = \overline{1,k}$. From here and from the condition 3) of statement 5 we arrive at $\mathfrak{R}_M = \sum_{i=1}^{k} \mathfrak{R}_{M(|S_i|,1)}$, and so any base $B \in \mathfrak{B}_M$ is of the form $B = \{a_1, a_2, \ldots, a_k\}$, where $a_i \in S_i, i = \overline{1,k}$. Suppose the bases $B_1 = \{a_1, a_2, \ldots, a_k\}$ and $B_2 = \{b_1, b_2, \ldots, b_k\}$. Then $B_1 \oplus B_2 = \sum_{i=1}^{k} (a_i \oplus b_i)$. As either $a_i \oplus b_i = \varnothing$ or $a_i \oplus b_i = \{a_i, b_i\} \in \mathfrak{R}_{M(|S_i|,1)}$, then $B_1 \oplus B_2 = \sum C$, $C \in \mathfrak{R}_M$, and the proof of necessity follows from theorem 15.□

If there are loops in the matroid M, i.e., $\mathfrak{I}_M(\varnothing) \neq \varnothing$, then the sets of loops in matroid M and H coincide. Obviously, $r_M(S - \mathfrak{I}_M(\varnothing)) = r_M(S)$ and, therefore, for the restriction $M|(S - \mathfrak{I}_M(\varnothing))$ it follows from the conditions 2) and 3) of statement 5 that $r_{M|(S-\mathfrak{I}_M(\varnothing))}(S - \mathfrak{I}_M(\varnothing)) = r_M(S - \mathfrak{I}_M(\varnothing)) = r_M(S)$ and $\mathfrak{R}_M = \mathfrak{R}_{M|(S-\mathfrak{I}_M(\varnothing))} + \mathfrak{I}_M(\varnothing)$.

From this we arrive at a generalization of theorem 19 for the case of arbitrary matroids $M \in \mathfrak{N}(S)$.

Theorem 20. *A top reduction mapping of a binary matroid* $M \in \mathfrak{N}(S)$ *of the rank* $r_M(S) = k$ *would be a* G*-mapping if and only if* $S - \mathfrak{I}_M(\varnothing) = S_1 + S_2 + ... + S_k$, *where* $|S_i| \geq 1$, $S_i \cap S_j = \varnothing$ *for all* $1 \leq i < j \leq k$ *and the equation* $M|(S - \mathfrak{I}_M(\varnothing)) = M(|S_1|, 1) + M(|S_2|, 1) + ... + M(|S_k|, 1)$ *holds true.*

For example, in the conditions of theorem 18 for matroid H^* the set of loops $\mathfrak{I}_{H^*}(\varnothing) = S - C$ and the lattice $\mathfrak{I}_{H^*|C}$ are isomorphic to the lattice of a free matroid of the rank 2, defined on the set C and representable as a sum of matroids $M(|D_0|, 1) + M(|C - D_0|, 1)$.

Theorem 21. *If an elementary mapping of binary matroids* $M \xrightarrow{\Phi} H$ *is, at the same time, a top reduction mapping and a* G*-mapping, then the mapping of dual matroids* $H^* \xrightarrow{\Phi^*} M^*$ *would also be a* G*-mapping.*

Proof. It follows from theorem 19 that for any co-point $K \in \mathfrak{K}_M$ the equation $K = S_{i_1} + S_{i_2} + ... + S_{i_{k-1}} + \mathfrak{I}_M(\varnothing)$, $1 \leq i_1 < i_2 < ... < i_{k-1} \leq k$ holds true. As $S = S_1 + S_2 + ... + S_k + \mathfrak{I}_M(\varnothing)$, we arrive at $\mathfrak{K}_M = \{S - S_i | i = \overline{1,k}\}$. Because the matroid H is the top reduction of the matroid M, similarly $\mathfrak{K}_H = \{S - S_i - S_j | 1 \leq i < j \leq k\}$. Hence, the respective families of cycles of dual matroids $\mathfrak{R}_{M^*} = \{S_i | i = \overline{1,k}\}$ and $\mathfrak{R}_{H^*} = \{S_i + S_j | 1 \leq i < j \leq k\}$. As $\mathfrak{R}_{M^*} \cap \mathfrak{R}_{H^*} = \varnothing$ and $\mathfrak{R}_{M^*} - \mathfrak{R}_{H^*} = \mathfrak{R}_{M^*}$, it follows that for any cycles

$C_1^*, C_2^* \in \mathfrak{R}_{M^*}$ the binary sum $C_1^* \oplus C_2^* = C^* \in \mathfrak{R}_{H^*}$. According to theorem 15, the mapping $H^* \xrightarrow{\Phi^*} M^*$ would be G-mapping.□

Consider the diagram:

$$\begin{array}{ccc} M & \xrightarrow{\Phi} & H \\ \updownarrow & & \updownarrow \\ M^* & \xleftarrow{\Phi^*} & H^*, \end{array} \tag{13}$$

where matroids M and H and their dual matroids M^* and H^* are binary and connected by elementary mappings, generated by the respective modular filters Φ_M and $\Phi_{H^*}^*$.

Theorem 22. *If in the diagram (13) the matroid H is a G-factor of the matroid M, then for any cycles $D \in \mathfrak{R}_H - \mathfrak{R}_M$ and $D^* \in \mathfrak{R}_{M^*} - \mathfrak{R}_{H^*}$ the equation $|D \cap D^*| = 2\mu + 1$, where $\mu \geq 0$, is valid.*

Proof. Suppose that G-factor $H = H(D_0)$ for some set $D_0 \in \mathfrak{F}_M, D_0 \neq \varnothing$, and consider a semi-matroid $\mathbb{R}(M, D_0)$. From theorems 2 and 6 it follows that the cycles of semi-matroids are from the family $\mathbb{R}_M(D_0) = \mathfrak{R}_{H(D_0)} - \mathfrak{R}_M$, and the bases are the co-points from the family $\mathbb{B}_M(D_0) = \mathfrak{K}_M - \mathfrak{K}_{H(D_0)}$. As for any base pseudo-matroids, and, therefore, for semi-matroids, the idempotency property is applicable, for any base $K \in \mathbb{B}_M(D_0)$ and element $a \in S - K$ there would always be found a single cycle $D(a, K) \in \mathbb{R}_M(D_0)$, such that $a \in D(a, K) \subseteq K \cup a$.

As with arbitrary binary matroids, for the base cycles of semi-matroid the condition (1) is valid, and therefore, for any cycle $D \in \mathbb{R}_M(D_0)$ and base $K \in \mathbb{B}_M(D_0)$, if $D - K = \{a_1, \ldots a_t\}$, then $D = D(a_1, K) \oplus \ldots \oplus D(a_t, K)$.

Let us note that $t \geq 1$, as cycles $\mathbb{R}_M(D_0)$ cannot be contained in bases $\mathbb{B}_M(D_0)$ as subset. If $t = 2\mu$, then the equation $D = (D(a_1, K) \oplus D(a_2, K)) \oplus \ldots \oplus (D(a_{t-1}, K) \oplus D(a_t, K))$ holds true, and, according to theorem 15, $D = \sum C, C \in \mathfrak{R}_M$. This contradicts the assumption that $D \in \mathbb{R}_M(D_0)$, and therefore $t = 2\mu + 1$, where $\mu \geq 0$.

On the other hand, as $K \in \mathfrak{K}_M - \mathfrak{K}_{H(D_0)}$ and $S - K = D^* \in \mathfrak{R}_{M^*} - \mathfrak{R}_{H^*(D_0)}$, according to the property 3) of statement 2, $\{a_1,...,a_t\} \subseteq D^*$, and, therefore, $|D \cap D^*| = |\{a_1,...a_t\}| = 2\mu + 1$, where $\mu \geq 0$. So, the proof of the statement is implied by the fact that the latter equation holds true for any cycles $D \in \mathfrak{R}_{H(D_0)} - \mathfrak{R}_M$ and any co-points $K \in \mathfrak{K}_M - \mathfrak{K}_{H(D_0)}$, and, therefore, for any cycles $D^* = S - K \in \mathfrak{R}_{M^*} - \mathfrak{R}_{H^*(D_0)}$.

2.3. Duality and *G*-Factorization

Apart from the conditions described in statement 4, there exist other conditions of binarity for matroids. In particular, it can be shown [12] that a matroid M would be binary if and only if $C_1 \oplus C_2 = C \in \mathfrak{R}_M$ for all cycles $C_1, C_2 \in \mathfrak{R}_M$, such that $C_1 \cap C_2 \neq \varnothing$, and the subsets $C_1, C_2 \subseteq S$ from a modular couple, i.e., $r_M(C_1 \cup C_2) = |C_1 \cup C_2| - 2.$

Theorem 23. *For any binary matroid* $M \in \mathfrak{N}(S)$ *the factorization of the mapping* $B(S) = M_0 \to M_1 \to ... \to M_k = M$ *is a G-factorization if and only if the factorization* $M^* = M_k^* \to M_{k-1}^* \to ... \to M_0^* = B^*(S)$ *of the respective dual matroid is also G-factorization.*

Proof. Consider the following diagram for some $i, 1 \leq i \leq k$, that is similar to (13)

$$\begin{array}{ccc} M_{i-1} & \longrightarrow & M_i \\ \updownarrow & & \updownarrow \\ M_{i-1}^* & \longleftarrow & M_i^* \end{array},$$

and fulfills the conditions of theorem 22. If $\mathfrak{R}_{M_{i-1}^*} \cap \mathfrak{R}_{M_i^*} = \varnothing$, then the diagram in question also fulfills the conditions of theorem 21 and, therefore, mapping $M_i^* \to M_{i-1}^*$ is a G-mapping.

Suppose $\mathfrak{R}_{M_{i-1}^*} \cap \mathfrak{R}_{M_i^*} \neq \varnothing$, cycles $D_1^*, D_2^* \in \mathfrak{R}_{M_{i-1}^*} - \mathfrak{R}_{M_i^*}$ and $D^* = D_1^* \oplus D_2^*$. Then, considering theorem 22, for any $D \in \mathfrak{R}_{M_i} - \mathfrak{R}_{M_{i-1}}$ the following equation holds true: $|D \cap D^*| = |D \cap D_1^*| + |D \cap D_2^*| - 2|D \cap D_1^* \cap D_2^*| = (2\mu_1 + 1) + (2\mu_2 + 1) - 2\mu$, where $\mu_1, \mu_2 \geq 0$ and $\mu = |D \cap D_1^* \cap D_2^*|$, i.e., $|D \cap D^*| = 2t, t \geq 0$. If D_1^*, D_2^* form a modular couple in matroid M_{i-1}^* and $D_1^* \cap D_2^* \neq \varnothing$, then the set D^* would be cycle and, according to the property 2) of statement 4 and considering the above, $D^* \in \mathfrak{R}_{M_{i-1}^*} \cap \mathfrak{R}_{M_i^*}$. If, on the contrary, the couple is not modular, then, according to the axiom of cycles for binary matroids, we get $D^* = \sum D_j^*$, where either $D_j^* \in \mathfrak{R}_{M_{i-1}^*} - \mathfrak{R}_{M_i^*}$ or $D_j^* \in \mathfrak{R}_{M_{i-1}^*} \cap \mathfrak{R}_{M_i^*}$, and the number of mutually non-intersection cycles $D_j^* \in \mathfrak{R}_{M_{i-1}^*} - \mathfrak{R}_{M_i^*}$ on the set D^* must be even. If all cycles D_j^* in the direct sum belong to the set $\mathfrak{R}_{M_{i-1}^*} \cap \mathfrak{R}_{M_i^*}$, then the statement follows from theorem 15.

Suppose there is a couple of mutually non-intersection cycles D_1^*, D_2^* that belong to the set $\mathfrak{R}_{M_{i-1}^*} - \mathfrak{R}_{M_i^*}$, such that $D_1^* + D_2^* \subseteq D^*$. Without loss of generality, let us consider the situation when $D^* = D_1^* + D_2^*$. It is obvious, then, that $D^* \in \Phi(\mathfrak{R}_{M_{i-1}^*} - \mathfrak{R}_{M_i^*})$, and, therefore, $r_{M_{i-1}^*}(D^*) = r_{M_i^*}(D^*) - 1$. As $r_{M_{i-1}^*}(D^*) \leq |D^*| - 2$, $r_{M_i^*}(D^*) \leq |D^*| - 1$. In other words, there exists a cycle $C^* \in \mathfrak{R}_{M_i^*} - \mathfrak{R}_{M_{i-1}^*}$, such that $C^* \subseteq D^*$. Moreover, for any element $a \in D^*$ the set $D^* - a$ also belongs to the modular filter $\Phi(\mathfrak{R}_{M_{i-1}^*} - \mathfrak{R}_{M_i^*})$, and $r_{M_{i-1}^*}(D^* - a) = r_{M_i^*}(D^* - a) - 1$. However, because $r_{M_{i-1}^*}(D^* - a) = r_{M_{i-1}^*}(D^*)$, $r_{M_i^*}(D^* - a) = r_{M_i^*}(D^*)$. Thus, the set D^* is a union of cycles in both matroid M_{i-1}^* and matroid M_i^*. Consequently, if D_1^*, D_2^* form a modular couple, i.e., $r_{M_{i-1}^*}(D^*) = |D^*| - 2$, then $r_{M_i^*}(D^*) = |D^*| - 1$ and

$D^* = C^* \in \mathfrak{R}_{M_i^*} - \mathfrak{R}_{M_{i-1}^*}$. If the couple D_1^*, D_2^* is not modular, then $C^* \subset D^*$ and, according to the theorem 3, $C^* = \overline{D_1^*} \cup \overline{D_2^*}$, where $\overline{D_1^*}, \overline{D_2^*} \in \mathfrak{R}_{M_{i-1}^*} - \mathfrak{R}_{M_i^*}$. So, there would always be found a cycle $\overline{D^*} \in \mathfrak{R}_{M_{i-1}^*} - \mathfrak{R}_{M_i^*}$, that is different from cycles D_1^* and D_2^*, and such that $D_1^* \cap \overline{D^*} \neq \varnothing$ and $D_2^* \cap \overline{D^*} \neq \varnothing$. This, in turn, means that $D^* = (D_1^* \oplus \overline{D^*}) + (D_2^* \oplus \overline{D^*})$, and the situation described above occurs repeatedly. If the pairs $D_1^*, \overline{D^*}$ and $D_2^*, \overline{D^*}$ are modular, then $D^* = C_1^* + C_2^*$ and the cycles $C_1^*, C_2^* \in \mathfrak{R}_{M_{i-1}^*} \cap \mathfrak{R}_{M_i^*}$. If, for instance, the pair $D_1^*, \overline{D^*}$ is not modular, then the binary sum $D_1^* \oplus \overline{D^*}$ analogously breaks down into a sum of non-intersection cycles, where the number of cycles from the family $\mathfrak{R}_{M_{i-1}^*} - \mathfrak{R}_{M_i^*}$ must be even. At that the sum of the respective couples of cycles would be the strict subset of the set D^*. In other words, iterative procedures of breakdown would inevitably lead to a situation where the respective couples of cycles would be modular.

Summarizing all that has been said, we arrive at $D^* = \sum C^*$, $C^* \in \mathfrak{R}_{M_i^*}$, and from theorem 15 we get that for any $i, 1 \le i \le k$, if the matroid M_i in the considered diagram is a G-factor of the matroid M_{i-1}, then the matroid M_{i-1}^* is also a G-factor of the matroid M_i^*. □

Let us illustrate these outcomes with an example using the same method as for semi-matroids.

Example 4. Suppose $S = \{1,2,3,4,5\}$. In example 3 a matroid M was considered with a family of cycles $\mathfrak{R}_M = \{\{1,2,3\},\{3,4,5\},\{1,2,4,5\}\}$. It is obvious that M is a binary matroid. Suppose $D_0 = \{\{1,2\}\} \in \mathfrak{F}_M$. According to (9), the family of cycles of the semi-matroids $\mathbb{R}(M, D_0)$ is the set $\mathbb{R}_M(D_0) = \{\{1,2\},\{3\},\{4,5\}\}$. Let us note, as (10) implies, that $\mathbb{R}_M(D_0) = \varnothing$, and the family of cycles $\mathfrak{R}_{H(D_0)}$ of the elementary factor $H(D_0)$

coincides with the set $\mathbb{R}_M(D_0)$. The lattices of flats $\mathfrak{I}_M$ and $\mathfrak{I}_{H(D_0)}$ are presented in Picture 3.

It is clear from Picture 3 that matroid $H(D_0)$ is an elementary factor of matroid M, generated by a modular cut $\mathfrak{Y}_M = \{\{3\},\{1,2,3\},\{3,4,5\},S\}$, and that the modular filter $\Phi_M = \Phi(\{1,2\},\{3\},\{4,5\}) = \Phi(\mathbb{R}_M(D_0))$.

It is easy to show that $\mathfrak{B}_{H(D_0)} = \{\{1,4\},\{1,5\},\{2,4\},\{2,5\}\}$ is the family of flats of matroid $H(D_0)$.

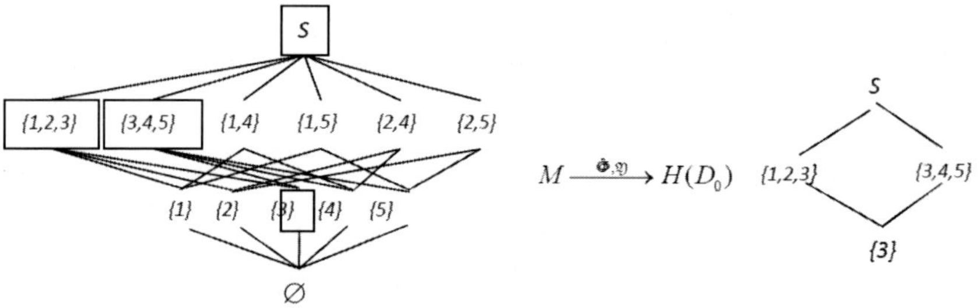

Picture 3. The lattices of flats $\mathfrak{I}_M$ and $\mathfrak{I}_{H(D_0)}$.

Let us highlight two points. As $\mathfrak{R}_M \cap \mathfrak{R}_{H(D_0)} = \mathfrak{R}_M(D_0) = \varnothing$, first of all, according to theorems 9 and 23, the mapping of the dual matroids $H^*(D_0) \xrightarrow{\Phi^*} M^*$ must also be a top reduction mapping, and G-mapping; second of all, according to theorems 6 and 7, the equation $\mathbb{B}_M(D_0) = \mathfrak{K}_M - \mathfrak{K}_{H(D_0)} = \mathfrak{B}_{H(D_0)}$ must hold true. Indeed, from Picture 3 we obtain that $\mathbb{B}_M(D_0) = \mathfrak{K}_M - \mathfrak{K}_{H(D_0)} = \{\{1,4\},\{1,5\},\{2,4\},\{2,5\}\} = \mathfrak{B}_{H(D_0)}$. Further, it is easy to show, for instance, using the property 3) from statement 2, that $\mathfrak{R}_{M^*} = \{\{1,2\},\{4,5\},\{1,3,4\},\{1,3,5\},\{2,3,4\},\{2,3,5\}\}$ and $\mathfrak{R}_{H^*(D_0)} = \{\{1,2\},\{4,5\}\}$ are the families of cycles of the dual matroids M^* and $H^*(D_0)$. Respectively, the lattices of flats $\mathfrak{I}_{M^*}$ and $\mathfrak{I}_{H^*(D_0)}$ are of the form (Picture 4):

Picture 4 demonstrates that the matroid M^* is indeed the top reduction of the matroid $H^*(D_0)$, generated by the modular cut $\mathfrak{Y}^*_{H^*} = S$ or the modular filter $\Phi^*_{H^*(D_0)} = \Phi(\mathfrak{B}_{H^*(D_0)})$. From the form of the family cycles $\mathfrak{R}_{H^*(D_0)}$ of matroid $H^*(D_0)$ we obtain that the family of its bases $\mathfrak{B}_{H^*(D_0)} = \{\{1,3,4\},\{1,3,5\},\{2,3,4\},\{2,3,5\}\}$.

Suppose $D_0^* = \{1,3,4\}$. Then, by definition,

$$\mathbb{R}_{H^*(D_0)}(D_0^*) = \{\{1,3,4\},\{1,3,5\},\{2,3,4\},\{2,3,5\}\} = \mathfrak{B}_{H^*(D_0)}$$

and $\mathfrak{R}_{H^*(D_0)}(D_0^*) = \{\{1,2\},\{4,5\}\} = \mathfrak{R}_{H^*(D_0)}$.

Thus, the matroid M^* simultaneously the top reduction and G-factor of matroid H^*, generated by the modular filter $\Phi^*_{H^*} = \Phi(\mathbb{R}_{H^*(D_0)}(D_0^*))$. Let us note that, according to the outcomes obtained earlier, we get both the equation $\mathfrak{R}_{H^*(D_0)} = \mathfrak{R}_{M^*} - \mathfrak{B}_{H^*(D_0)}$ and $\mathfrak{R}_{M^*} - \mathfrak{R}_{H^*(D_0)} = \mathfrak{B}_{H^*(D_0)}$. For any cycles $D \in \mathbb{R}_M(D_0)$ and $D^* \in \mathbb{R}_{H^*(D_0)}(D_0^*)$ the equation $|D \cap D^*| = 1$ holds true, according to theorem 22. Further, the dual matroids M^* and $H^*(D_0)$ must fulfill theorem 19. According to Picture 4, the lattice of flats $\mathfrak{I}_{H^*(D_0)}$ is isomorphic to the lattice of flats of a free matroid of the rank 3, the set $S = \{1,2\} + \{3\} + \{4,5\}$ and matroid $H^*(D_0)$ is representable as a sum of matroids: $M(|\{1,2\}|,1) + M(|\{3\}|,1) + M(|\{4,5\}|,1)$, which is in accordance with theorem 19.□

For any binary matroid M the dual matroid M^* is also binary, and it is always possible to construct a G-factorization $M^* = M_k^* \to M_{k-1}^* \to \ldots \to M_0^* = B^*(S)$. Thus, theorem 23 implies that for all binary matroids $M \in \mathfrak{N}(S)$ there exists a G- factorization $B(S) = M_0 \to M_1 \to \ldots \to M_k = M$ of the canonical mapping $B(S) \to M$. Hence, by means of G-factorization all binary matroids from the set $\mathfrak{N}(S)$ are enumerated.

Let us note that, as the dual to a binary matroid is also binary, $|\mathfrak{N}_k(S)| = |\mathfrak{N}_{|S|-k}(S)|$.

Statement 22. *The number of binary matroids of the rank k in the set of all the binary matroids* $\mathfrak{N}(S)$ *fulfills the equation*

$$|\mathfrak{N}_k(S)| = |\mathfrak{N}_{|S|-k}(S)| = \frac{(2^{|S|}-1)(2^{|S|-1}-1)...(2^{|S|-k+1}-1)}{(2-1)(2^2-1)...(2^k-1)}. \tag{14}$$

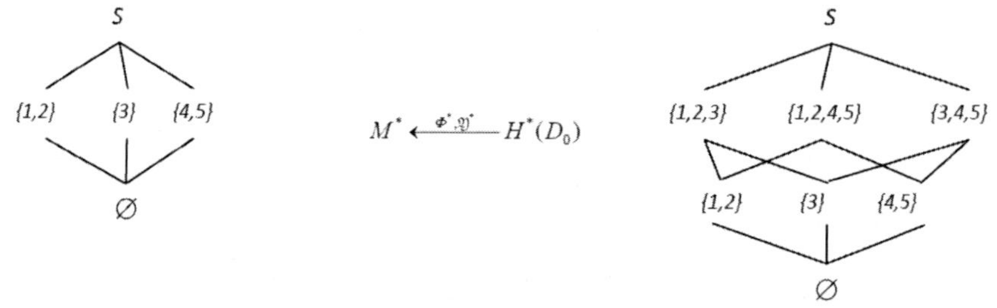

Picture 4. The lattices of flats $\mathfrak{I}_{M^*}$ and $\mathfrak{I}_{H^*(D_0)}$.

Proof. A corollary of theorem 16 suggests that $|\mathfrak{N}_{|S|-1}(S)| = |\mathfrak{N}_1(S)| = 2^{|S|} - 1$. Let us prove the formula (14) by induction.

Suppose that in diagram (13) matroid $M \in \mathfrak{N}_{|S|-k}(S)$, and matroid $H \in \mathfrak{N}_{|S|-k-1}(S)$ is its *G*-factor. From the properties 3) and 4) of theorem 12 it follows, firstly, that all $2^{|S|-k} - 1$ different non-empty subsets of any base $B \in \mathfrak{B}_M$ generate different *G*-factors, i.e., different matroids $H \in \mathfrak{N}_{|S|-k-1}(S)$, and, secondly, that different bases produce the same family of *G*-factors. On the other hand, for a fixed matroid $H \in \mathfrak{N}_{|S|-k-1}(S)$ the dual matroid $H^* \in \mathfrak{N}_{k+1}(S)$ analogously generates $2^{k+1} - 1$ different matroids $M^* \in \mathfrak{N}_k(S)$ and, therefore, the same number of different matroids

$M \in \mathfrak{N}_{|S|-k}(S)$, for which the matroid H would be the same G-factor. Therefore, $|\mathfrak{N}_{|S|-k-1}(S)| = |\mathfrak{N}_{|S|-k}(S)| \dfrac{2^{|S|-k}-1}{2^{k+1}-1}$, which concludes the proof. □

Binary matroids from the set $\mathfrak{N}_k(S)$ are isomorphic to vector matroids and represent able by binary matrixes order $k \times |S|$ and rank k. Vector subspaces of rank k in the vector subspaces of rank $|S|$ correspond to every such matrix. Thus, the number of different matroids from the set $\mathfrak{N}_k(S)$ coincides with the number of such different vector subspaces of rank k, which is in accordance with formula (14).

These outcomes enable us, by means of a step-by-step algorithm of G-factorizations, to construct the families of cycles of all the different matroids from the set $\mathfrak{N}_k(S)$ for any $k, 1 \leq k \leq |S|$. Let us note that this way the matroids are constructed as the respective families of cycles explicitly and not the binary matrixes, by means of which vector matroids can be constructed, but which are not matroids themselves.

So, the outcomes of this section reveal that a family of semi-matroids, generated by G-mappings, can be thought of as algebraic objects, axiomatically defined by a set of binary matroids $\mathfrak{N}(S)$. At that, any binary matroid M, according to theorem 23, is representable as a result of G-factorization of a canonical mapping $B(S) \to M$ and, therefore, as a sequence of respective semi-matroids. Hence, and from statement 19, matroids from the set $\mathfrak{N}(S)$ can be viewed as objects of a category, closed under morphisms – G-mappings and rank-preserving weak mappings.

In weak order on the set of all matroids $\mathfrak{M}(S)$ simple mappings lower the rank of the initial matroid if and only if they are top reduction mappings. Therefore, theorem 20 describes the conditions in which the weak order covers of binary matroids from the set $\mathfrak{N}(S)$ which lower the rank of the initial matroid would not only be top reductions, but also G-mappings.

***

In the second part of this work the type of pseudo-matroids, generated by elementary mappings of matroids and, as a consequence, functionally connected to the category of matroids and their mappings, is introduced into the theory of matroids. A pseudo-matroid unequivocally describes the elementary mapping by which it is generated, and so every matroid $M \in \mathfrak{M}(S)$ is representable as a sequence of the respective pseudo-matroids for any factorization of a canonical mapping $B(S) \to M$. Cycles of pseudo-matroids are the minimal by inclusion elements of modular filters, and bases are the co-points of matroid *M,* which are different from the co-points of the respective modular cuts. In the investigation of properties of pseudo-matroids, this fact enables us to obtain some outcomes, which, in turn, make it possible to view pseudo-matroids, generated by the elementary mappings of matroids, as an independent algebraic notion in the general theory of matroids.

In the second section a new notion of binarity preserving the *G*-mappings of binary matroids is introduced. Semi-matroids are considered as a special case of pseudo-matroids, generated by *G*-mappings. The main outcome of the section is the proof of the fact that the category of binary matroids is closed under morphisms - *G*-mappings and rank-preserving weak mappings. It follows from this that any binary matroid $H \in \mathfrak{N}(S)$ is presentable as a sequence of semi-matroids, generated specifically by some *G*-factorization of a canonical mapping $B(S) \to H$. It is proven that the set of semi-matroids, generated by all *G*-factors of any binary matroid, can be defined as an algebraic structure, axiomatically set by matroids from the set $\mathfrak{N}(S)$.

Chapter 3

ENUMERATION OF ALL NON-ISOMORPHIC MATROIDS

In this part the problem of enumeration for all non-isomorphic matroids is considered for both the general case of the set $\mathfrak{M}(S)$ and the specific case of binary matroids from the set $\mathfrak{N}(S)$.

Let us denote the number of all non-isomorphic matroids, defined on the n-element set $S, |S| = n$, by *f(n)*. In [12] an estimate is given

$$n - \frac{3}{2}\log n + O(\log\log n) \le \log\log f(n) \le n - \log n + O(\log\log n) ,(15)$$

from which it can be inferred that the total number of matroids in the set $\mathfrak{M}(S)$ is estimated by a value of order $n!2^{\frac{1}{n}2^n}$.

Thus, the practical implementability of any enumerative algorithm of all non-isomorphic matroids actually depends on both the number of non-isomorphic matroids and on the existence of a non-enumerative ("constructive") procedure of their extraction from the set $\mathfrak{M}(S)$.

At the moment of this work's creation, the results of the enumeration of all non-isomorphic matroids for $1 \le |S| \le 8$, obtained by way of continuous calculations [12], were known. The problem of their enumeration for $|S| > 8$ still remained unsolved.

The third part is thematically divided into two sections.

The first section introduces the notion of an "isomorphism of pseudo-matroids" for the general case of matroids from the set $\mathfrak{M}(S)$ and describes a procedure of a constructive check of the isomorphism of matroids, defined by their families of cycles.

The second section considers a method of solution of the enumeration problem for all non-isomorphic binary matroids from the set $\mathfrak{N}(S)$ and provides an algorithm of construction of their families of cycles the computational complexity of $O(n^3 3^n)$ arithmetical operations. The respective outcomes for the values $1 \le n \le 15$ are presented.

1. Non-Isomorphic Matroids

1.1. Isomorphism of Matroids and Pseudo-Matroids

An isomorphism of matroids $M_1, M_2 \in \mathfrak{M}(S)$ of the same rank, from the point of view of families of their cycles, means that there exists a one-to-one correspondence π on the set *S*, such that for any cycle $C = (c_1, \ldots, c_k) \in \mathfrak{R}_{M_1}$ the set $\pi(C) = (\pi(c_1), \ldots, \pi(c_k))$ is a cycle of matroid M_2.

For any matroid *M* there exists a factorization of the canonical mapping $B(S) \to M$ into elementary mappings

$$B(S) = M_0 \to M_1 \to \ldots \to M_{|S| - r_M(S)} = M \tag{16}$$

and, therefore, every matroid *M* is representable by a sequence of pseudo-matroids $\mathbb{R}(M_{i-1}, M_i), 1 \le i \le |S| - r_M(S)$. On the other hand, for every matroid *M* among the factorizations (16) there would be a Higgs factorization. At that, according to theorem 8, the family of cycles of Higgs lifts is produced by a constructive mechanism. However, the matroid *M* itself may not be a Higgs lift of any other matroid, as its dual matroid *M**, according to theorem 10, would be erectable and, therefore, not just any matroid from the set $\mathfrak{M}(S)$. Therefore, to construct all matroids of rank $r_M(S)$ it is necessary to be able to construct all possible sequences of pseudo-matroids $\mathbb{R}(M_{i-1}, M_i), 1 \le i \le |S| - r_M(S)$.

Definition 29. Let us call pseudo-matroids $\mathbb{R}(M_1,H_1)$ and $\mathbb{R}(M_2,H_2)$ *isomorphic,* if the matroids M_1 and M_2 would be isomorphic, and so would their elementary factors H_1 and H_2.

If π is a one-to-one correspondence on the set *S,* then we shall denote the matroid that is isomorphic to the matroid *M* by $\pi(M)$.

Statement 23. *For any elementary factor H of matroid M pseudo-matroids* $\mathbb{R}(M,H)$ *and* $\mathbb{R}(\pi(M),\pi(H))$ *are isomorphic.*

Proof. Suppose that the elementary factor *H* of the matroid *M* is generated by the modular filter $\Phi_M = \Phi(\mathfrak{R}_H - \mathfrak{R}_M)$. Let us show, then, that the matroids $\pi(M)$ and $\pi(H)$ would also be connected by an elementary mapping, generated by the modular filter $\Phi_{\pi(M)} = \Phi(\pi(\mathfrak{R}_H) - \pi(\mathfrak{R}_M))$. Indeed, considering the fact that the correspondence π is one-to-one, the equation $\pi(\mathfrak{R}_H - \mathfrak{R}_M) = \pi(\mathfrak{R}_H) - \pi(\mathfrak{R}_M)$ is valid, and, considering also the properties of isomorphic matroids, if a cycle *D* belongs to the set $\mathfrak{R}_H - \mathfrak{R}_M$, then the cycle $\pi(D)$ belongs to the set $\pi(\mathfrak{R}_H) - \pi(\mathfrak{R}_M)$. As for isomorphic of elementary factors matroids (uniquely corresponds to each flat $\overline{A} \in \mathfrak{I}_H$) the flat $\pi(\overline{A}) \in \mathfrak{I}_{\pi(H)}$, $r_H(\overline{A}) = r_{\pi(H)}(\pi(\overline{A}))$ and from the modularity of the filter Φ_M of the matroid *M* follows the modularity of the filter $\Phi_{\pi(M)}$ of matroid $\pi(M)$.

This means that for the enumeration of all non-isomorphic matroids in the set $\mathfrak{M}(S)$ it is sufficient, firstly, to construct all possible pseudo-matroids as ordered couples $< S, \mathbb{R}_{M_{i-1}}(M_i) >, 1 \le i \le |S| - r_M(S)$ on every step of factorization (16), and, secondly, to have a procedure for singling out all non-isomorphic matroids from these, to then use them at the next step of the factorization.

In this connection, let us remind ourselves that the lattice of all elementary factors of any matroid *M,* ordered by weak order, is isomorphic to the lattice of all its modular filters, ordered by inclusion. Let us show that the pseudo-matroids $\mathbb{R}(M,H_1)$ and $\mathbb{R}(M,H_2)$ are non-isomorphic, if the elementary factors H_1 and H_2 of matroid *M* are comparable in weak order.

Indeed, suppose the factors H_1 and H_2 are generated by the modular filter Φ_1 and Φ_2. Then, if $\Phi_1 \subset \Phi_2$, $\mathfrak{R}_{H_1} - \mathfrak{R}_M \subset \mathfrak{R}_{H_2} - \mathfrak{R}_M$ and $\mathfrak{B}_{H_2} \subset \mathfrak{B}_{H_1}$. Hence, there would not exist a one-to-one correspondence π on the set *S*, which transfers the families of cycles and bases of matroids H_1 and H_2 into one another.

Thus, it only makes sense to check the pseudo-matroids, constructed in each step of factorization (16), isomorphism, if either the respective elementary factors of some matroid are incomparable in weak order or the pseudo-matroids are generated as a result of factorization of some non-isomorphic matroids.

The family of cycles $\mathfrak{R}_M$ of an arbitrary matroid *M*, like any other family of subsets, is unequivocally described by its *incidence matrix* or a dual graph $\Gamma(S, \mathfrak{R}_M)$, the nodes of which would be the elements of the set *S* and the cycles from the family $\mathfrak{R}_M$. An element $a \in S$ is *incident* or connected in graph $\Gamma(S, \mathfrak{R}_M)$ with a cycle $C \in \mathfrak{R}_M$ by an edge, if $a \in C$. So, the matroids M_1 and M_2 are isomorphic if and only if the respective graphs $\Gamma(S, \mathfrak{R}_{M_1})$ and $\Gamma(S, \mathfrak{R}_{M_2})$ are isomorphic. It follows from this that in isomorphic matroids not only the total number of cycles of certain length coincides, but also the number of cycles incident to any element $a \in S$. At the same time, this condition is necessary, but not sufficient for the isomorphism of matroids.

For any matroid M let us denote

$$\mathfrak{R}_M(r, A) = \{C \in \mathfrak{R}_M \,|\, A \subseteq C, |C| = r\}, 1 \le r \le r_M(S) + 1$$

Then it is obvious that the subset $\mathfrak{R}_M(r) = \{C \in \mathfrak{R}_M \,\|\, C| = r\}$ of all cycles of the length *r* in $\mathfrak{R}_M$ coincides with the set $\bigcup_{a \in S} \mathfrak{R}_M(r, a)$.

For any subset $A \subseteq S$ let construct an integer vector

$$\Omega_M(A) = \left(|\mathfrak{R}_M(1, A)|, |\mathfrak{R}_M(2, A)|, \ldots, |\mathfrak{R}_M(r_M(S) + 1, A)| \right). \tag{17}$$

Suppose $W_M(A) = \sum_{r=1}^{r_M(S)+1} r\,|\mathfrak{R}_M(r,A)|$.

Clearly, the vector $\Omega_M(A)$ can be a zero vector, and, therefore, $W_M(A) = 0$, for instance, for subsets *A*, such that $|A| > r_M(S)+1$.

For any $r, 1 \le r \le r_M(S)+1$ the number of cycles of the length *r* in the family $\mathfrak{R}_M$ is given by the equality $|\mathfrak{R}_M(r)| = \frac{1}{r}\sum_{a\in S}|\mathfrak{R}_M(r,a)|$, while $\sum_{r=1}^{r_M(S)+1} |\mathfrak{R}_M(r,a)|$ is the total number of cycles, incident with the node $a \in S$ in the graph $\Gamma(S, \mathfrak{R}_M)$.

As has already been mentioned, if a one-to-one correspondence π on the set *S* generates an isomorphism of matroids M_1 and M_2, then $|\mathfrak{R}_{M_1}(r)| = |\mathfrak{R}_{M_2}(r)|$ for all possible values *r* and $\sum_{r=1}^{r(S)+1} |\mathfrak{R}_{M_1}(r,a)| = \sum_{r=1}^{r(S)+1} |\mathfrak{R}_{M_2}(r,\pi(a))|$ for all elements $a \in S$, where $r(S) = r_{M_1}(S) = r_{M_2}(S)$. However, these conditions alone are not sufficient for the isomorphism of matroids M_1 and M_1. In particular, the total number of different elements $a \in S$, that belong to cycles from the families $\mathfrak{R}_{M_1}(r)$ and $\mathfrak{R}_{M_2}(r)$, must coincide, i.e., the equation $|\{a \in S \mid a \in C, C \in \mathfrak{R}_{M_1}(r)\}| = |\{b \in S \mid b \in D, D \in \mathfrak{R}_{M_2}(r)\}|$ must hold true for all values *r*, $1 \le r \le r(S)+1$, and also both the number of connectivity components and the number of different elements in them must coincide too.

Statement 24. *Matroids* $M_1, M_2 \in \mathfrak{M}(S)$ *of rank* $r(S)$ *are isomorphic if and only if there exists a one-to-one correspondence* π *on the set S, such that* $\Omega_{M_1}(A) = \Omega_{M_2}(\pi(A))$ *any subset* $A \subseteq S$.

Proof. For isomorphic matroids M_1 and M_1 the equality of vectors (17) is obvious.

Suppose $C \in \mathfrak{R}_{M_1}(r)$ for some $r, 1 \le r \le r(S)+1$. Because cycles cannot be contained within each other as subsets, it follows from $\Omega_{M_1}(C) = \Omega_{M_2}(\pi(C))$, that $|\mathfrak{R}_{M_1}(r,C)| = |\mathfrak{R}_{M_2}(r,\pi(C))| = 1$. Hence, considering the definition of the set $\mathfrak{R}_{M_2}(r,\pi(C))$, $\pi(C)\in\mathfrak{R}_{M_2}(r)$. The isomorphism of matroids of the same rank, therefore, follows from the fact that this inclusion is valid for any cycles $C\in\mathfrak{R}_{M_1}(r)$ and for $r, 1\le r\le r(S)+1$.□

1.2. The Algorithm of Isomorphism Check for Matroids from the Set $\mathfrak{M}(S)$

For any subset $A\subseteq S$ let us denote by $A(m)=\{B\subseteq A \,\|\, |B|=m\}$ the family of all its subsets of cardinality $m, 1\le m\le |A|$. Considering this, now suppose $\Omega_{M_1}{}^{(m)}=\{\Omega_{M_1}(A)\,|\,A\in S(m)\}$ and $\Omega_{M_2}{}^{(m)}=\{\Omega_{M_2}(B)\,|\,B\in S(m)\}$. We shall call the families of integer vectors $\Omega_{M_1}{}^{(m)}$ and $\Omega_{M_2}{}^{(m)}$ *equivalent,* if they coincide up to the rearrangement of the element of the set *S*, and denote them by $\Omega_{M_1}{}^{(m)}\sim\Omega_{M_2}{}^{(m)}$. The equivalent condition means the fulfillment of the two following conditions.

First, the family of subsets $S(m)$ is divided into classes of equivalence

$$S(m)=S_{M_1}^{(1)}(m)+...+S_{M_1}^{(t_m)}(m)=S_{M_2}^{(1)}(m)+...+S_{M_2}^{(t_m)}(m) \tag{18}$$

so that up to numeration, $|S_{M_1}^{(i)}(m)|=|S_{M_2}^{(i)}(m)|$ for all $i, 1\le i\le t_m$, and $\Omega_{M_1}(A)=\Omega_{M_2}(B)$ for any subsets $A,B\in S(m)$ if and only if $A\in S_{M_1}^{(i)}(m)$ and $B\in S_{M_2}^{(i)}(m)$ for some $i, 1\le i\le t_m$.

Second, there exists a permutation of elements π_m on the set *S*, such tha, if $A\in S_{M_1}^{(i)}(m)$, then $\pi_m(A)\in S_{M_2}^{(i)}(m)$ for all $i, 1\le i\le t_m$.

The permutations π_{m_1} and π_{m_2}, $m_1\le m_2$ we shall call *non-contradictory* as one-to-one correspondences on the set *S*, if for any set $A\in S(m_2)$ from the condition $B\in A(m_1)$ follows, that $\pi_{m_1}(B)\subseteq\pi_{m_2}(A)$.

Statement 25. *Matroids* $M_1, M_2 \in \mathfrak{M}(S)$ *of rank* $r(S)$ *are isomorphic if and only if* $\Omega_{M_1}{}^{(m)} \sim \Omega_{M_2}{}^{(m)}$ *for all* $m, 1 \le m \le r(S)$, *and the respective permutations* π_m *are non-contradictory as one-to-one correspondences on the set S.*

Proof. It is sufficient to show that on the set of elements of *S* there exists a permutation π, such that $\Omega_{M_1}(A) = \Omega_{M_2}(\pi(A))$ for all subsets $A \subseteq S$, $|A| \le r(S)$. In the conditions of the statement this fact follows from the non-contradiction of permutations π_m, defined on the set *S*, by induction.

Let us highlight that, by definition, the permutations π_m would be non-contradictory to the permutations π_{m-1} only if the equivalence classes the form (18) are considered not on the whole set *S*, but on the subsets, for which there exist permutations π_{m-1} that fulfill the conditions of statement 25.

Based on statement 25, a constructive algorithm of isomorphism check for matroids M_1 and M_2 of the same rank, defined their families of cycles $\mathfrak{R}_{M_1}$ and $\mathfrak{R}_{M_2}$, can be suggested.

For $m = 1$ it is obvious that $S(1) = S$, and the divided (18) is of the form

$$S = S_{M_1}^{(1)} + ... + S_{M_1}^{(t_1)} = S_{M_2}^{(1)} + ... + S_{M_2}^{(t_1)} \qquad (19)$$

As in the divided (19) the sets of elements that belong to different classes do not intersect, the graphs $\Gamma(S, \mathfrak{R}_{M_1})$ and $\Gamma(S, \mathfrak{R}_{M_2})$ would be isomorphic if and only if the subgraphs $\Gamma(S_{M_1}^{(i)}, \mathfrak{R}_{M_1})$ and $\Gamma(S_{M_2}^{(i)}, \mathfrak{R}_{M_2})$ would be isomorphic for all $i, 1 \le i \le t_1$. In other words, any permutation π_1 that fulfills the conditions of statement 25 can be presented the following way: $\pi_1 = (\pi_1^{(1)}, ..., \pi_1^{(t_1)})$, where $\pi_1^{(i)} : S_{M_1}^{(i)} \leftrightarrow S_{M_2}^{(i)}, i = \overline{1, t_1}$, are permutations, defined on the subsets of the set *S*, respective to the divided (19).

Thus, for $m = 2$ of permutation π_2 that fulfill the conditions of statement 25 would only exist if $\{\Omega_{M_1}(A) \mid A \in S_{M_1}^{(i)}(2)\} \sim \{\Omega_{M_2}(B) \mid B \in S_{M_2}^{(i)}(2)\}$ for all i, $1 \le i \le t_1$. The respective divided would be of the form:

$$S_{M_1}^{(i)}(2) = S_{M_1}^{(i,1)}(2) + ... + S_{M_1}^{(i,t_2(i))}(2) \text{ and } S_{M_2}^{(i)}(2) = S_{M_2}^{(i,1)}(2) + ... + S_{M_2}^{(i,t_2(i))}(2) \quad (20)$$

It is obvious that not only the number of bigrams in the respective divided classes (20), but also the number of different elements in them, must coincide, i.e., the permutations π_2 must be such that $\{a \in A \mid A \in S_{M_1}^{(i,j)}(2)\} \leftrightarrow \{b \in B \mid B \in S_{M_2}^{(i,j)}(2)\}$ for all $i, 1 \le i \le t_1$ and $j, 1 \le j \le t_2(i)$. In this case, the permutations π_1 and π_2 are non-contradictory only if for any $i, 1 \le i \le t_1$ and any element $a \in S_{M_1}^{(i)}$ the following holds true $|\{A \in S_{M_1}^{(i,j)}(2) \mid a \in A\}| = |\{B \in S_{M_2}^{(i,j)}(2) \mid \pi_1^{(i)}(a) \in B\}|, j = \overline{1, t_2(i)}$.

This implies that in the general case the permutation $\pi_1 = (\pi_1^{(1)}, ..., \pi_1^{(t_1)})$ breaks down into $\sum_{i=1}^{t_1} t_2(i)$ permutations, so that

$$\pi_2 = (\pi_1^{(1,1)}, ..., \pi_1^{(1,t_2(1))}, ..., \pi_1^{(t_1,1)}, ..., \pi_1^{(t_1,t_2(t_1))}). \quad (21)$$

Therefore, as with $m = 1$, the graphs $\Gamma(S, \mathfrak{R}_{M_1})$ and $\Gamma(S, \mathfrak{R}_{M_2})$ are isomorphic only if the subgraphs, defined on subsets of the set S, which correspond to the permutations $\pi_1^{(i,j)}$, where $i = \overline{1, t_1}$ and $j = \overline{1, t_2(i)}$, are isomorphic.

Obviously, if the graphs $\Gamma(S, \mathfrak{R}_{M_1})$ and $\Gamma(S, \mathfrak{R}_{M_2})$ are isomorphic, then the permutations π_m are non-contradictory for all $m, 1 \le m \le r(S)$. Otherwise, either the families of vectors $\Omega_{M_1}^{(m)}$ and $\Omega_{M_2}^{(m)}$ are not equivalent, or the permutations π_{m-1} and π_m are contradictory for some m.

Let us point out that it only makes sense to this algorithm to construct the permutations π_m for such values of m for which the subsets $A \in S_{M_1}^{(i,j)}(m)$ and

$A \in S_{M_2}^{(i,j)}(m)$ exist for some $i, 1 \le i \le t_{m-1}$ and $j, 1 \le j \le t_m(i)$, such that $W_{M_1}(A) \ne 0$ or $W_{M_2}(A) \ne 0$ respectively. For example, if the number of different elements in all classes of divided (20) does not exceed two, then the isomorphism check ends on the value $m = 2$, regardless of the value of $r(S)$. At the same time, the connectivity components of matroids M_1 and M_2, the number of which and the number of different elements in which must coincide for isomorphic matroids, are completely defined by the equivalence classes (20). Provided that these parameters coincide, for identifying an isomorphism of matroids M_1 and M_2, it would be possible to confine the values $m = 2$ also for the general case, checking the condition $\pi_2(C) \in \mathfrak{R}_{M_2|S''}$ for all cycles $C \in \mathfrak{R}_{M_1|S'}$ and couples of connectivity components S' and S'', such that $\pi_2\left(S'\right) = S''$.

This way, if it were possible to construct the families of cycles and all elementary factors for any non-isomorphic matroid M_{i-1} in factorization (16) by means of some non-enumerative procedure, then it would be possible to list all non-isomorphic matroids $M_i, 1 \le i \le |S| - r_M(S)$ using the constructive algorithm, presented above.

As has been noted above, it only makes sense to compare the values of coordinates of vectors $\Omega_{M_1}(a)$ and $\Omega_{M_2}(b)$ for constructing the equivalence classes (19) only for such elements $a, b \in S$, for which $W_{M_1}(a) = W_{M_2}(b)$. An analogous situation occurs for the subsets $A \in S(2)$ in the divided (20). Thus, the number of equivalence checks of families of vectors $\Omega_{M_1}^{(m)} \sim \Omega_{M_2}^{(m)}$ for $m = 1, 2$ substantially depends only on the value of the sum $t_1 + \sum_{i=1}^{t_1} t_2(i)$. In this respect, let us highlight that the care of $t_1 = 1$ in (19) for connected matroids only occurs when the matroids M_1 and M_2 are in fact uniform matroids $M(|S|, r(S))$, which are isomorphic by definition for any rank $r(S), 1 \le r(S) \le |S|$ and, therefore, they would fulfill the conditions of statement 25 for all $m, 1 \le m \le r(S)$. At the same time, $\left|\mathfrak{R}_{M(|S|, r(S))}(r, a)\right| = 0$, if

$r \neq r(S)+1$ and $\left|\mathfrak{R}_{M(|S|,r(S))}(r(S)+1,a)\right| = \binom{|S|}{r(S)}$ for all elements $a \in S$.

So, the uniformity of matroids M_1 and M_2 is easily identified by the form of vectors $\Omega_{M_1}(a)$ and $\Omega_{M_2}(a), a \in S$, and therefore, the question of their isomorphism for $m=1$ is answered. The same situation occurs for the connectivity components of separable matroids. For arbitrary matroids, therefore, $t_1 > 1$ for $m=2$, and on average $t_1 + \sum_{i=1}^{t_1} t_2(i) = \mathrm{O}\left(\frac{|S|^2}{2t_1}\right)$.

As a conclusion, it is important to point out that if there is a non-enumerative procedure of isomorphism check for pseudo-matroids, the computational complexity of the problem of enumeration of all non-isomorphic matroids from the set $\mathfrak{M}(S)$ is determined not only by their quantity, but also by the current absence of a constructive algorithm for building the families of cycles of all elementary factors of an arbitrary matroid, acceptable by complexity. In particular, also for that reason, it took persistent complex calculations to list all non-isomorphic matroids, defined on the *set* $S, |S|=8$, the number of which, according to the estimate (15), does not exceed 2^{32}. In the next section we shall look into the problem of the enumeration of all non-isomorphic binary matroids from the set $\mathfrak{N}(S)$ and show that there occurs a principally different situation if G-factorization of canonical mappings is used.

2. Non-Isomorphic Binary Matroids

As opposed to the general case, the number of different binary matroids in the set $\mathfrak{N}(S)$ is known exactly, specifically $|\mathfrak{N}(S)| = \sum_{k=1}^{|S|} |\mathfrak{N}_k(S)|$, where $|\mathfrak{N}_k(S)|$ is calculated by formula (14). At the same time, at the moment of re-edition of Oxley's monograph [12] in 2006, the number of non-isomorphic matroids, as with the set $\mathfrak{M}(S)$, was known only for $1 \leq |S| \leq 8$. Additionally, for binary matroids an asymptotic result was proven: for $|S| \to \infty$ the number of non-isomorphic binary matroids on the set $\mathfrak{N}(S)$ tends to $\frac{1}{|S|!}|\mathfrak{N}(S)|$.

Let us point out that, as in the general case, the enumeration of binary matroids means, for example, the construction of their families of cycles. Further, a method of solution of this problem, based on G-factorization of canonical mappings, will be considered.

2.1. *G*-Lifts of Binary Matroids

Definition 30. If a matroid H is a G-factor of a matroid $M \in \mathfrak{N}(S)$, then we shall call the matroid M a G-*lift* of the matroid H.

It follows from theorem 23 that any binary matroid M can be obtained as a result of not just a factorization, but a G-factorization of a canonical mapping $B(S) \to M$. At that, all G-factors on every step $i, 1 \le i \le |S| - r_M(S)$, of factorization (16) are constructed, as described in part II, by means of iterations, and the computational complexity is estimated by the value of $O(2^{|S|-r_{M_{i-1}}(S)})$. Let us show that an analogous result can be obtained for all G-lifts of binary matroids. Let us note that for arbitrary matroids, based on theorem 8, a constructive procedure for building the families of cycles of lifts of canonical mappings of Higgs lifts alone can be suggested. In this respect, let us remind ourselves that, according to definition 30, a G-lift of a binary matroid $H \in \mathfrak{N}(S)$ may not be its lift under a canonical mapping $B(S) \to M$ in the classical meaning of definition 20.

Any matroid $N \in \mathfrak{M}(S)$ and a base $B \in \mathfrak{B}_N$ have an unequivocally corresponding family of fundamental cycles

$$\mathfrak{C}_N(B) = \{C_N(b^*, B) \in \mathfrak{R}_N \,|\, b^* \in S - B\}. \quad (22)$$

Theorem 24. *In the conditions of diagram (13) for any base* $B_0 \in \mathfrak{B}_H$ *of the subset* $D_0^* \subseteq B_0^* = S - B_0 \in \mathfrak{B}_{H^*}$, $D_0^* \neq \varnothing$, *such that the modular filter*

$\Phi^*_{H^*} = \Phi(\mathbb{R}_{H^*}(D_0^*))$, *and an element* $d_0^* \in D_0^*$, *the set* $B_0 \cup d_0^* \in \mathfrak{B}_M$ *and the respective family of fundamental cycles is of the form*

$$\mathfrak{C}_M(B_0 \cup d_0^*) = \{C_H(b^*, B_0) \in \mathfrak{R}_H \,|\, b^* \in B_0^* - D_0^*\} +$$

$$+\{C_H(b^*, B_0) \oplus C_H(d_0^*, B_0) \,|\, b^* \in D_0^* - d_0^*\}. \quad (23)$$

Proof. Consider the family of fundamental cycles $\mathfrak{C}_H(B_0) = \{C_H(b^*, B_0) \in \mathfrak{R}_H \,|\, b^* \in B_0^*\}$ of the matroid H. As, by assumption, the matroid M^* in scheme (13) is a G-factor of the matroid H^* and the modular filter $\Phi^*_{H^*} = \Phi(\mathbb{R}_{H^*}(D_0^*))$, where the subset $D_0^* \subseteq B_0^*$, $D_0^* \neq \varnothing$, then $|C_H(b^*, B_0) \cap D_0^*| = 1$, if $b^* \in D_0^*$, and $|C_H(b^*, B_0) \cap D_0^*| = 0$, if $b^* \in B_0^* - D_0^*$. In the conditions of the statement and according to scheme (13), by definition, the matroid M would be a G-lift of the matroid H. Hence, considering theorem 22, $C_H(b^*, B_0) \in \mathfrak{F}_M$, if $b^* \in D_0^*$ and $C_H(b^*, B_0) \in \mathfrak{R}_M \cap \mathfrak{R}_H$, and, therefore, $C_H(b^*, B_0) \in \mathfrak{R}_M$, if $b^* \in B_0^* - D_0^*$ which represents the first summand in (23).

This also implies that $C_H(d_0^*, B_0) \in \mathfrak{F}_M$ for any element $d_0^* \in D_0^*$. And, as $d_0^* \notin B_0$, the set $B_0 \cup d_0^* \in \mathfrak{F}_M$, and also, as $r_M(S) = r_H(S) + 1$, the set $B_0 \cup d_0^*$ belongs to the family $\mathfrak{B}_M$. Suppose $d^* \in D_0^* - d_0^*$. Then there exists a single fundamental cycle $C_M(d^*, B_0 \cup d_0^*)$ in the matroid M for the element d^* in the base $B_0 \cup d_0^*$. At that, it is also necessary that $d_0^* \in C_M(d^*, B_0 \cup d_0^*)$, as otherwise the equation $C_M(d^*, B_0 \cup d_0^*) = C_H(d^*, B_0)$ would take place, which contradicts the condition $C_H(d^*, B_0) \in \mathfrak{F}_M$. As the cycles $C_H(d^*, B_0)$ and $C_H(d_0^*, B_0)$ are fundamental, either their binary sum $C_H(d_0^*, B_0) \oplus C_H(d^*, B_0)$ is a cycle in the set $\mathfrak{R}_H$, or the equation $C_H(d_0^*, B_0) \oplus C_H(d^*, B_0) = C_H(d_0^*, B_0) + C_H(d^*, B_0)$ holds true. Otherwise, we get $|(C_H(d_0^*, B_0) \oplus C_H(d^*, B_0)) \cap D_0^*| = 2$. Therefore, according to theorem 22 and considering the uniqueness of fundamental cycles, either $C_M(d^*,B_0 \cup d_0^*) = C_H(d_0^*,B_0) \oplus C_H(d^*,B_0) \in \mathfrak{R}_M \cap \mathfrak{R}_H$ or $C_M(d^*,B_0 \cup d_0^*) = C_H(d_0^*,B_0) + C_H(d^*,B_0) \in \mathfrak{R}_M - \mathfrak{R}_H \cap \mathfrak{R}_M$.□

For any base, including the base $B_0 \cup d_0^* \in \mathfrak{B}_M$, the set of cycles $\mathfrak{R}_M$ of a binary matroid M is defined by the family of fundamental cycles the following way: $\mathfrak{R}_M = \{C = \sum \oplus D \,|\, D \in \mathfrak{C}_M(B_0 \cup d_0^*)$ and $C - \min\}$.

Let us note that, according to (23), in diagram (13) the pseudo-matroid $\mathbb{R}(M,H)$, regarded as a semi-matroid, is defined by an ordered couple $< S, \mathbb{R}_M(C_H(d_0^*,B_0)) >$ in the conditions of theorem 24, and is generated by the cycle $C_H(d_0^*,B_0) \in \mathfrak{R}_H$.

According to theorem 23, if a matroid $M \in \mathfrak{N}(S)$ is a G-lift of a matroid $H \in \mathfrak{N}(S)$, then it would also be a G-lift in the G-factorization of a canonical mapping $B(S) \to H$. As $\mathfrak{R}_M \cap \mathfrak{R}_H \neq \varnothing$, it follows from theorem 8 that for any binary matroid $H \in \mathfrak{N}(S)$ of rank $r_H(S) \le |S| - 2$ the G-factorization of a canonical mapping $B(S) \to H$ would not be a Higgs factorization at the same time. However, as the matroid H is an elementary factor of the matroid M, as has already been pointed out, the matroid M would also be its Higgs lift in matroid M. So, it follows from what has been said above that the matroid $M \in \mathfrak{N}(S)$, being a Higgs lift in the matroid M for any of its G-factors H, would not be its Higgs lift in the matroid $B(S)$, i.e., under the canonical mapping of $B(S) \to H$, as opposed to G-factorization.

From theorem 24, therefore, we get an algorithm for the construction of all G-lifts of any given binary matroid H.

1. For an arbitrary base $B_0 \in \mathfrak{B}_H$ we get a base $B_0^* = S - B_0 \in \mathfrak{B}_{H^*}$.
2. For all subsets $D^* \subseteq B_0^*$, $D^* \neq \varnothing$, choose an element $d_0^* \in D^*$ and construct the families of fundamental cycles $\mathfrak{C}_M(B_0 \cup d_0^*)$ according to (23).
3. Given the families of fundamental cycles, we get the sets of cycles of respective G-lifts M of matroid H:

$$\mathfrak{R}_M = \{C = \sum \oplus D \big| D \in \mathfrak{C}_M(B_0 \cup d_0^*) \text{ and } C - \min\}.$$

Let us illustrate this algorithm with an example.

Example 5. Consider the matroids from example 4 and the diagram

$$\begin{array}{ccc} M & \xrightarrow{D_0} & H(D_0) \\ \updownarrow & & \updownarrow \\ M^* & \xleftarrow{D_0^*} & H^*(D_0) \end{array}.$$

Suppose in theorem 24 the base $B_0 \in \mathfrak{B}_{H(D_0)}$ and $B_0 = \{2,5\}$. Then $B_0^* = \{1,3,4\} \in \mathfrak{B}_{H^*(D_0)}$. The fundamental cycles of the matroid $H(D_0)$ for the base B_0 are of the form: $C_{H(D_0)}(\{1\}, B_0) = \{1,2\}$, $C_{H(D_0)}(\{3\}, B_0) = \{3\}$ and $C_{H(D_0)}(\{4\}, B_0) = \{4,5\}$. Suppose that the base $B \in \mathfrak{B}_M$ and $B = \{1,2,5\}$, so that $B^* = \{3,4\} \in \mathfrak{B}_{M^*}$. The respective fundamental cycles of the matroid M for the base B: $C_M(\{3\}, B) = \{1,2,3\}$ and $C_M(\{4\}, B) = \{1,2,4,5\}$.

In example 4 the set $D_0^* = \{1,3,4\}$. Suppose $d_0^* = \{1\}$. As $B_0^* = D_0^*$, according to (23), $\mathfrak{C}_M(\{2,5\} + \{1\}) = \{C_{H(D_0)}(\{3\}, B_0) \oplus C_{H(D_0)}(\{1\}, B_0)$, $C_{H(D_0)}(\{4\}, B_0) \oplus C_{H(D_0)}(\{1\}, B_0)\} = \{\{1,2,3\}, \{1,2,4,5\}\}$. Thus, the family $\mathfrak{C}_M(\{2,5\} + \{1\})$, defined in 24, coincides with the family of fundamental cycles of the matroid M for the base $B = \{1,2,5\}$, and the family of cycles, based on it, coincides with the set $\mathfrak{R}_M = \{\{1,2,3\}, \{1,2,4,5\}, \{3,4,5\}\}$.

As a conclusion, let us point out that by means of this algorithm the families of cycles of G-lifts of binary matroids, as well as those of G-factors, can be built constructively.

It was shown before that this situation does not occur for matroids from the set $\mathfrak{M}(S)$. At the same time, similar to G-factorization, the factorization of a canonical mapping $B(S) \to H$, $H \in \mathfrak{M}(S)$ would be a Higgs factorization if and only if the factorization of the respective mapping of dual matroids $H^* \to B^*(S)$ would also be a Higgs factorization. However, if the families of cycles of Higgs lifts of a canonical mapping $B(S) \to H$ can be built constructively, then the families of elementary factors of the dual matroid H^*, for which the matroid H^* would be a Higgs lift, cannot.

2.2 The Algorithm for Enumerating All Non-Isomorphic Binary Matroids

According to the outcomes obtained by Lucas [9], any rank-preserving weak mapping of binary matroids increases the number of connectivity components. Thus, as for arbitrary matroids, the G-factors of a binary matroid, comparable in weak order, are non-isomorphic. Let us remind ourselves that the information about the number of connectivity components becomes accessible when equivalence classes (20) are built for $m=2$.

Statement 26. *For the identification of the isomorphism of matroids* $M_1, M_2 \in \mathfrak{N}(S)$ *of the rank* $r(S)$ *it is sufficient to check the fulfillment of the conditions of statement 25 for* $m=1,2$.

Proof. Suppose that on the set S a permutation π_2 is set, such that $\Omega_{M_1}(A)=\Omega_{M_2}(\pi_2(A))$ for all subsets $A \subseteq S$, for which $|A| \leq 2$. Then $\left|\mathfrak{R}_{M_1}(r)\right| = \left|\mathfrak{R}_{M_2}(r)\right|$ for any $r, 1 \leq r \leq r(S)+1$, and the number of connectivity components, as well as the number of different elements in them, are the same for matroids M_1 and M_2. In particular, the number of loops coincides, i.e., $\left|\mathfrak{R}_{M_1}(1)\right| = \left|\mathfrak{R}_{M_2}(1)\right|$, and any one-to-one correspondence between them would be a component of π_2, defined on the set S.

Let us denote $h_r = \left|\mathfrak{R}_{M_1}(r)\right| = \left|\mathfrak{R}_{M_2}(r)\right|$, $2 \leq r \leq r(S)+1$, and suppose $\mathfrak{R}_{M_1}(r) = \{C_1, C_2, \ldots, C_{h_r}\}$. We shall prove the statement inductively on h_r.

Suppose that $h_r = 1$ and $a \in C_1$. Then $\left|\mathfrak{R}_{M_1}(r,a)\right| = \left|\mathfrak{R}_{M_2}(r,\pi_2(a))\right| = 1$ and for any subset $D \in S(2)$, such that $a \in D$ we get $\left|\mathfrak{R}_{M_1}(r,D)\right| = \left|\mathfrak{R}_{M_2}(r,\pi_2(D))\right| = 1$, if $D \subseteq C_1$, and $\left|\mathfrak{R}_{M_1}(r,D)\right| = \left|\mathfrak{R}_{M_2}(r,\pi_2(D))\right| = 0$, if $D \not\subseteq C_1$. Considering the equation $|\pi_2(C_1)| = r$, we arrive at $\pi_2(C_1) \in \mathfrak{R}_{M_2}(r)$. Let us denote $\mathfrak{R}^{(h)}_{M_1}(r) = \{C_1, C_2, \ldots, C_h\}$. Suppose that $\pi_2(C) \in \mathfrak{R}_{M_2}(r)$ for any cycle $C \in \mathfrak{R}^{(h)}_{M_1}(r)$ and $h_r = h+1$. For the cycle C_{h+1} there are two alternative options: either $C_{h+1} \not\subset \bigcup_{i=1}^{h} C_i$ or $C_{h+1} \subset \bigcup_{i=1}^{h} C_i$. In the first case there would always be found an element $a \in S$, such that $a \in C_{h+1}$, $a \notin \bigcup_{i=1}^{h} C_i$, and, therefore, $\left|\mathfrak{R}_{M_1}(r,a)\right| = \left|\mathfrak{R}_{M_2}(r,\pi_2(a))\right| = 1$. It is easy to show that this situation corresponds to the case of $h_r = 1$, considered above, and thus $\pi_2(C_{h+1}) \in \mathfrak{R}_{M_2}(r)$. If the inclusion $C_{h+1} \subset \bigcup_{i=1}^{h} C_i$ takes place, then again two possibilities exist. Either for some $l, 2 \le l \le h$ the equation $C_{h+1} = C_{i_1} \oplus C_{i_2} \oplus \ldots \oplus C_{i_l}$ is valid, where $1 \le i_1 < i_2 < \ldots < i_l \le h$, or it is not valid for all $l, 2 \le l \le h$. In the first case $\pi_2(C_{h+1}) = \pi_2(C_{i_1}) \oplus \pi_2(C_{i_2}) \oplus \ldots \oplus \pi_2(C_{i_l})$, and as $\left|\pi_2(C_{h+1})\right| = r$, by the axiom of cycles for binary matroids, $\pi_2(C_{h+1}) \in \mathfrak{R}_{M_2}(r)$. In the latter case for any element $a \in C_{h+1}$ and subset $D \in S(2)$, for which $a \in D$, $\left|\mathfrak{R}_{M_1}(r,a)\right| = \left|\mathfrak{R}_{M_2}(r,\pi_2(a))\right| = \left|\mathfrak{R}_{M_1}^{(h)}(r,a)\right| + 1$, $\left|\mathfrak{R}_{M_1}(r,D)\right| = \left|\mathfrak{R}_{M_2}(r,\pi_2(D))\right| = \left|\mathfrak{R}_{M_1}^{(h)}(r,D)\right| + 1$, if $D \subseteq C_{h+1}$, and $\left|\mathfrak{R}_{M_1}(r,D)\right| = \left|\mathfrak{R}_{M_2}(r,\pi_2(D))\right| = \left|\mathfrak{R}_{M_1}^{(h)}(r,D)\right|$, if $D \not\subseteq C_{h+1}$, respectively. In other words, considering the induction assumption, the situation is analogous to that when $h_r = 1$, which was analyzed earlier, and therefore $\pi_2(C_{h+1}) \in \mathfrak{R}_{M_2}(r)$. □

It follows from statement 26 that, to check the isomorphism of binary matroids $M_1, M_2 \in \mathfrak{N}(S)$ when using the algorithm described in the first section, it is sufficient to consider the values of the parameter $m = 1,2$.

Let us now consider an algorithm for building the families of cycles of all non-isomorphic binary matroids from the set $\mathfrak{N}(S)$, based on G-factorizations of canonical mappings. In the suggested algorithm the families of cycles of non-isomorphic matroids are built consequently, according to the matroid's rank. This fact enables us to use the known outcomes of enumeration [12], which were obtained for $1 \le |S| \le 8$. In particular, according to them, the number of non-isomorphic binary matroids of rank $|S| - k$ can be estimated by the value of $O\left(\binom{|S|}{k}\right)$.

Theorem 25. *There exists an algorithm for building the families of cycles of all non-isomorphic binary matroids from the set* $\mathfrak{N}(S)$ *of computational complexity* $O(|S|^3 3^{|S|})$, *which is based on the procedure of* $\mathfrak{G}$*-factorization of canonical mappings.*

Proof. Suppose $S=\{1,2,...,n\}$, a binary matroid $M\in\mathfrak{N}_{n-k}(S)$, and a G-factorization of a canonical mapping $B(S)\to M$ is of the form

$$B(S)=M_0\to M_1\to ...M_{k-1}\to M_k=M\,. \tag{24}$$

To build the families of cycles of all non-isomorphic matroids M_k of rank $n-k$, using the procedure of G-factorization (24), let us consider an algorithm that consists of the following stages.

1. For every matroid M_{k-1} from the array of non-isomorphic matroids of rank $n-k+1$, $2^{n-k+1}-1$ subsets $D_k\subseteq B$, $D_k\neq\varnothing$, are set, where the base $B\in\mathfrak{B}_{M_{k-1}}$, and the families of cycles of matroids M_k are constructed as $\mathfrak{G}$-factors of non-isomorphic matroids M_{k-1}, generated by modular filters $\Phi(\mathbb{R}_{M_{k-1}}(D_k))$.
2. From the family of cycles of the G-factor M_k vectors $\{\Omega_{M_k}(a)\,|a\in S\}$ are built, and the values of parameters $|\mathfrak{R}_{M_k}(r)|, 1\le r\le n-k+1$, and the numbers $W_{M_k}(a), a\in S$, are calculated.
3. The values of the parameters and numbers, obtained at stage 2, are compared with the already known analogous values for all the non-isomorphic G-factors M_k. If there are no coincidences, then the families of cycles, along with the respective vectors $\{\Omega_{M_k}(a)\,|a\in S\}$ and the values of the parameters and numbers $W_{M_k}(a), a\in S$, mentioned earlier are included into the array of non-isomorphic G-factors M_k.
4. If the values of parameters and numbers for some couples of matroids coincide at stage 3, then the algorithm for the isomorphism check for $m=1,2$, described in the previous section, is used. If the matroids are

non-isomorphic, then the comparisons of stage 3 are carried on for other non-isomorphic G-factors M_k, which were built earlier. If, as a result of all such checks, a matroid M_k is considered non-isomorphic, then the procedure of its inclusion into the array of non-isomorphic G-factors is repeated.

According to statements 23, 24 and 26, as a result of this algorithm the families of cycles of all non-isomorphic matroids M_k of rank $n-k$ would be built.

The main contribution to the computational complexity of this algorithm is made by stage 4, which is concerned with a comparison of $(n-k+1)$ numbers to check the equivalence of the respective families of vectors $\Omega_{M_k}^{(m)}$. It has already been pointed out that the number of equivalence checks for $m=1,2$ substantially depends only on the value of the sum $t_1+\sum_{i=1}^{t_1} t_2(i)=\mathrm{O}\left(\frac{n^2}{2t_1}\right)$. Thus, the number of the comparisons, mentioned above, can be estimated by $O\left(n^2\right)$.

Therefore, supposing that stage 4 is carried out for all non-isomorphic matroids M_{k-1}, we conclude that the total computational complexity of building the families of cycles of all non-isomorphic binary matroids of rank $n-k$, defined on the set S, $|S|=n$, does not exceed $O\left(n^2(n-k+1)\binom{n}{k-1}2^{n-k+1}\right)$ arithmetical operations. Accordingly, the computational complexity of building the families of cycles of all non-isomorphic binary matroids from the set $\mathfrak{N}(S)$ is estimated by $O(|S|^3 3^{|S|})$. □

This estimate allows an inference that, by means of this algorithm, based on the procedure of G-factorization of canonical mappings, the families of cycles of all non-isomorphic binary matroids from the set $\mathfrak{N}(S)$ can be built constructively. So the problem of their enumeration for the set S of big enough cardinality can be solved.

In Table 1 the results of calculations, based on the above mentioned algorithm, are given for all non-isomorphic binary matroids from the set $\mathfrak{N}(S)$ for $|S|=n$ and $n\leq 15$.

Table 1. The number of non-isomorphic binary matroids, defined on the set $S = \{1, 2, \ldots, n\}$ and of rank $k = \overline{1, n}$

k/n	1	2	3	4	5	6	7	8	9	10	11	12	13	14	15
0	1	1	1	1	1	1	1	1	1	1	1	1	1	1	1
1	1	2	3	4	5	6	7	8	9	10	11	12	13	14	15
2		1	3	6	10	16	23	32	43	56	71	89	109	132	158
3			1	4	10	22	43	77	131	213	333	507	751	1088	1546
4				1	5	16	43	106	240	516	1060	2108	4064	7641	14036
5					1	6	23	77	240	705	1988	5468	14724	39006	101818
6						1	7	32	131	516	1988	7664	29765	117169	467266
7							1	8	43	213	1060	5468	29765	173035	1074527
8								1	9	56	333	2108	14724	117169	1074527
9									1	10	71	507	4064	39006	467266
10										1	11	89	751	7641	101818
11											1	12	109	1088	14036
12												1	13	132	1546
13													1	14	158
14														1	15
15															1
Sum	2	4	8	16	32	68	148	342	848	2297	6928	24034	98854	503137	3318734

Let us note that, as the matroid, dual to a binary matroid, is also binary, the number of non-isomorphic matroids of ranks k and $n-k$ must coincide. At the same time, let us point out that the abovementioned algorithm consequently builds the families of cycles of non-isomorphic matroids M_k of rank $n-k$, based on the already known families of cycles of non-isomorphic matroids M_{k-1} of rank $n-k+1$, regardless of their potential duality to the already built matroids. In other words, the coincidence of the respective values in Table 1 is nothing but an extra validity check of this algorithm's realization.

***

As a conclusion, let us make a few notes.

In part III the problem of the enumeration of all non-isomorphic matroids from the sets $\mathfrak{M}(S)$ and $\mathfrak{N}(S)$ has been considered. At the time of this work's publication, the outcomes of enumeration for both the general case and the case of binary matroids were only known for $1 \le |S| \le 8$. The problem of their enumeration for $|S| > 8$ remained unsolved.

In the first section the notion of the "isomorphism of pseudo-matroids" was introduced, and a constructive algorithm for an isomorphism check for matroids from the set $\mathfrak{M}(S)$, under the condition that the families of their cycles are known, was offered. This allows us to bring the problem of enumeration down to the problem of construction or families of cycles of all elementary factors of an arbitrary matroid from the set $\mathfrak{M}(S)$. Provided there is such a procedure, the complexity of the enumeration problem solution for all non-isomorphic matroids from the set $\mathfrak{M}(S)$ would mainly depend on their quantity, which is estimated by a value of $2^{\frac{1}{|S|}2^{|S|}}$.

In part II it was proven that all the binary matroids $M \in \mathfrak{N}(S)$ can be built by means of *G*-factorization of canonical mappings. As opposed to the general case, the families of cycles of all $2^{r_M(S)}-1$ *G*-factors of any binary matroid are built constructively. Based on this fact, in the second section an algorithm was suggested for enumerating all non-isomorphic binary matroids from the set $\mathfrak{N}(S)$. The computational complexity of this algorithm for building the families of cycles of all non-isomorphic binary matroids from the set $\mathfrak{N}(S)$ does not exceed $O\left(|S|^3 3^{|S|}\right)$ arithmetical operations, which allows us to solve the problem of enumeration for a wide enough range of values of $|S|$. The outcomes for $1 \le |S| \le 15$ were presented.

It is important to highlight that this outcome confirms the validity of the previously established hypothesis of existence of exponential complexity algorithms for the enumeration of all non-isomorphic binary matroids.

Chapter 4

G-CODES AND THEIR PRACTICAL APPLICATIONS

In this part, the outcomes obtained earlier are applied to some practical problems.

The first section is dedicated to the problem of building all non-isomorphic optimal linear codes. In the second section a matroid interpretation of the classical problem of linear codes decoding is presented.

A new way of coding (*G*-coding), which generalizes the traditional convolutional encoding, is described in the third section. The fourth section analyzes a range of questions related to the noise-resistance of *G*-codes, and, based on the outcomes of the first section, an algorithm for building the most noise-resistant *G*-codes is offered.

1. NON-ISOMORPHIC OPTIMAL LINEAR CODES

In this section the results obtained earlier will be implemented in building all the non-isomorphic linear codes for quite a wide range of parameter values.

Let us denote the space of all binary vectors of length n by V_n. It is easy to show that a linear dependence between vectors generates a matroid $M(V_n)$ on the set V_n. This matroid can be considered as an analogue of a free matroid $B(S)$ defined on the set $S, |S| = n$. The analogy shows that if any set $A \subseteq S$ is a flat of rank $r_{B(S)}(A) = |A|$, where $1 \leq |A| \leq n$, in the matroid

$B(S)$, then any vector subspaces of rank $k, 1 \le k \le n$ would be flats in the matroid $M(V_n)$. In particular the co-points $M(V_n)$, or hyper planes - the vector subspaces of rank $n-1$. Let us remind ourselves that from this and from statement 1 it follows that for any vector subspace $X \subseteq V_n$ of rank k there would always be found $n-k$ hyper planes $H_1, H_2, ..., H_{n-k}$, such that $X = H_1 \cap H_2 \cap ... \cap H_{n-k}$.

A vector subspace X of rank k is called a linear (n,k)-code with the distance $d+1$, if $\chi(v) \ge d+1$ for all vectors $v \in X$, where $\chi(v)$ is Hamming weight; of a binary vector $v \in V_n$. An *optimal* [14] code is a code for which, for given values of parameters n and d, the value of parameter k is maximal. Such a value of parameter k of an optimal code we shall denote by $\psi(n,d)$. Let us point out that determining the parameter $\psi(n,d)$ is one of the central problems of algebraic coding theory.

In this context, for the subset of vectors $V(n,d) = \{v \in V_n \mid \chi(v) \le d\} \subseteq V_n$ the critical problem [14] lies in finding the minimal number of $h(n,d)$ hyper planes, the intersection of which does not contain any vectors from the set $V(n,d)$. Let us demonstrate that the critical problem in this case is closely related to the abovementioned coding problem.

Statement 27. The values $\psi(n,d)$ and $h(n,d)$, connected by the following formula:

$$\psi(n,d) + h(n,d) = n. \tag{25}$$

Proof. Suppose $H_1, H_2, ..., H_{h(n,d)}$ is the minimal set of hyper planes in V_n, such that $H_1 \cap H_2 \cap ... \cap H_{h(n,d)} \cap V(n,d) = \varnothing$. Then the set of vectors $X = H_1 \cap H_2 \cap ... \cap H_{h(n,d)}$ is a vector subspace of rank $n - h(n,d)$. Hence, by definition, $n - h(n,d) \le \psi(n,d)$, and, therefore, $h(n,d) \ge n - \psi(n,d)$.

Conversely, suppose Y is a subspace in V_n of rank $\psi(n,d)$, such that $Y \cap V(n,d) = \varnothing$. Then there would be found $n-\psi(n,d)$ hyper planes $H_1, H_2, \ldots, H_{n-\psi(n,d)}$, for which $Y = H_1 \cap H_2 \cap \ldots \cap H_{n-\psi(n,d)}$. Thus, by definition, we arrive at the equation $h(n,d) \le n - \psi(n,d)$, which, considering what has just been said, proves the equation (25). $\square$

Consider a matroid M. For a given integer parameter $d, d \ge 1$, the subset $A \subseteq S$ is called d-independent in the matroid M if:

1. $|A| \ge d$
2. $B \subseteq A, |B| = d \Rightarrow B \in \mathfrak{F}_M$.

Suppose that $|S| = n$, and let us define a subfamily of subset $S(n,d) = \{A \subseteq S \,\|\, A| \le d\} \subseteq 2^S$. Under these terms the following statement holds true.

Statement 28. The problem of the construction of optimal codes is identical to the problem of finding binary matroids $M \in \mathfrak{N}(S)$ of rank $r_M(S) = \psi(n,d)$, such that the set S would be d-independent in dual matroids $M^* \in \mathfrak{N}(S)$.

Proof. Suppose for some binary matroid M the set S would be d-dependent in the dual matroid M^*. If $|A| \le d$, then $A \in \mathfrak{F}_{M^*}$ and $S(n,d) \cap \mathfrak{R}_{M^*} = \varnothing$. Therefore, for any cycle $D \in \mathfrak{R}_{M^*}$ the inequation $|D| \ge d+1$ is valid. Suppose $D_1, D_2 \in \mathfrak{R}_{M^*}$ and $D_1 \ne D_2$. By the axiom of cycles for binary matroids, $D_1 \oplus D_2 = \sum D, D \in \mathfrak{R}_{M^*}$. As in the latter sum the cycles do not overlap, $|D_1 \oplus D_2| \ge d+1$. Therefore, the equation $\{\sum \oplus D \,|\, D \in \mathfrak{R}_{M^*}\} \cap S(n,d) = \varnothing$ holds true. The binary matroid M is isomorphic to the vector matroids, generated by a $r_M(S) \times n$ binary matrix, the non-zero elements rows of which are cycles from the set $\mathfrak{R}_{M^*}$. Thus, it follows from the latter equation that the linear code, built based on this matrix, would be a $(n, r_M(S))$-code with the distance of $d+1$. If $r_M(S) = \psi(n,d)$, then this code would be optimal.

Table 2. The number of non-isomorphic matroids for different values of $|S|, 1 \le |S| \le 22$, where the length of cycles of the dual matroids is not less than $d+1$, $3 \le d \le 8$

d = 3	**6**	**7**	**8**	**9**	**10**	**11**	**12**	**13**	**14**
2	1	3	7	13	21	31	44	59	77
3			4	12	33	74	150	275	471
4				5	24	92	311	902	2331
5					4	34	249	1429	6840
6						2	43	623	7342
7							2	47	1535
8								1	49
9									1
d = 4	**8**	**9**	**10**	**11**	**12**	**13**	**14**	**15**	**16**
2	1	4	9	16	26	38	53	71	92
3			2	11	36	91	195	373	654
4				1	14	92	424	1490	4328
5						15	282	2921	19669
6							11	1011	31010
7								6	4019
8									1
d = 5	**9**	**10**	**11**	**12**	**13**	**14**	**15**	**16**	**17**
2	1	3	7	14	23	35	50	68	89
3			1	7	24	67	156	320	592
4				1	7	47	248	1031	3484
5						6	98	1249	11859
6							5	185	8575
7								3	368
8									1
d = 6	**11**	**12**	**13**	**14**	**15**	**16**	**17**	**18**	**19**
2	1	4	9	17	28	42	59	80	104
3			1	7	27	80	194	408	773
4				1	7	48	307	1497	5669
5					1	7	95	1901	26431
6							3	113	19043
7								2	84
8									1

d = 7	**12**	**13**	**14**	**15**	**16**	**17**	**18**	**19**	**20**
2	1	3	7	14	24	37	54	74	98
3			1	4	16	50	133	301	611
4				1	5	23	131	688	3091
5					1	5	76	450	6907
6							2	30	1855
7								1	27
8									1
d = 8	**14**	**15**	**16**	**17**	**18**	**19**	**20**	**21**	**22**
2	1	4	9	17	29	44	63	86	113
3				2	14	53	151	361	749
4						7	91	735	4070
5							3	183	8005
6									248

In the second section of part III an algorithm for building the families of cycles of all non-isomorphic matroids from the set $\mathfrak{M}(S)$ has been offered. Let us note that this algorithm allows for building the families of cycles of all non-isomorphic binary matroids, for which there is a limitation on length. For this, it is enough to limit the stages 1) and 2) to matroids M_{k-1} and the sizes of subsets that generate their G-factors, which fulfill certain limitations. Clearly, this approach enables us to build the families of cycles of special non-isomorphic binary matroids for the set S of a rather big cardinality due to its smaller computational complexity. Thus, considering statement 28, the problem of constructing the families of cycles of all non-isomorphic optimal linear codes is solved for a wide range of values of parameters n,k and d.

Table 2 offers the results of the calculation of the number of non-isomorphic binary matroids for different values of $|S|, |S| \leq 22$, where the length of cycles of the dual matroids is not less than $d+1$, $3 \leq d \leq 8$.

For the given values of parameters $|S|, 1 \leq |S| \leq 22$ and d, $1 \leq d \leq 8$, shown in Table 2, we can obtain not only the values of parameter $\psi(|S|, d)$, but also the number of all the respective non-isomorphic optimal linear codes. Let us note that the realization of the algorithm from part III enables us to also build the generating matrixes for all non-isomorphic optimal linear codes.

2. (x, A, δ) SYSTEM AND (J, M, D^*)-SCHEMES

In this section the matroid interpretation of a range of problems concerned with decoding processes and with solving systems linear equations with skewed right-hand sides is presented.

Suppose A is a $k \times n$ matrix over the field $GF(2)$. The matrix A generates a subspace of rank k or a linear (n,k)-code $Y(A) = \{xA | x \in V_k\}$ space V_n. At the same time, it can be considered to be the generating matrix in the system of equations, where $Y(A)$ would be the set of right-hand sides of the equations of this system.

For any vector $y \in Y(A)$ there exists $2^{k-rangA}$ alternate vectors $x \in V_k$ that fulfill the equation $y = xA$. In the conditions described above the vector $x \in V_k$ is unequivocally defined.

Let us note that, in the general case, for arbitrary generating matrixes of systems of linear equations the situation $k \geq n$ и and, therefore, rang $A \leq k$, is possible. As has already been pointed out, this only affects the number of vectors $x \in V_k$ that fulfill equation $y = xA$, and does not result in any limitation of the generality of the problems considered in this section.

Suppose the vector $\delta \in V_n$ and $Y(A,\delta) = \{xA \overline{\oplus} \delta | x \in V_k\}$, where $\overline{\oplus}$ is a coordinate-wise sum of vectors over the field $GF(2)$. As above, the matrix A and the vector δ generate a system of linear equations with skewed right-hand sides, and $Y(A,\delta)$ is the set of its right-hand sides. From now on we shall call such systems (x, A, δ)-*systems*.

Consider the equation

$$z_0 = x_0 A \overline{\oplus} \delta_0, \tag{26}$$

where the matrix A and the vector $z_0 \in V_n$ are known. Obviously, $z_0 \in Y(A, \delta_0)$. The vector $\delta_0 \in V_n$ from (26) is called the *error vector*, and so $\Delta(z_0) = \{\delta = \delta_0 \overline{\oplus} xA | x \in V_k\}$ is the set of all possible error vectors in the respective (x_0, A, δ_0)-systems. It follows from (26) that $\Delta(z_0) = \{\delta = \delta_0 \overline{\oplus} xA | x \in V_k\}$, and for any possible error vector $\delta \in \Delta(z_0)$ there exists a vector $x \in V_k$ for which $z_0 = xA \overline{\oplus} \delta$. So, as opposed to the situation described earlier, for any fixed vector $z_0 \in V_n$, all vectors $x \in V_k$ fulfill the equation (26) a priori. Therefore, it is only possible to unequivocally determine an unknown vector $\delta_0 \in V_n$ from a (x_0, A, δ_0)-system, when the set of all possible error vectors $\Delta(z_0)$ is ordered and in this order there will be found unique minimal and maximal elements.

Let us denote the subset of all possible error vectors $\Delta_{\min}(z_0) \subseteq \Delta(z_0)$, and a vector $\delta_{\min} \in \Delta(z_0)$ belongs to the set $\Delta_{\min}(z_0)$ if and only if the condition $\chi(\delta_{\min}) = \min\limits_{\delta \in \Delta(z_0)} \{\chi(\delta)\}$ is fulfilled. If $|\Delta_{\min}(z_0)| = 1$, then there exists a unique vector $\delta_{\min} \in V_n$ for which $z_0 = x_{\min} A \overline{\oplus} \delta_{\min}$, and at that, if $\delta_{\min} = \delta_0$, then $x_{\min} = x_0$.

For example, suppose $Y(A)$ is a linear (n, k)-code with the distance $d + 1$ and $\chi(\delta_0) \le d/2$. Then $|\Delta_{\min}(z_0)| = 1$ and $\delta_{\min} = \delta_0$. In the general case, when $|\Delta_{\min}(z_0)| > 1$, it is only possible to distinguish the vector $\delta_{\min}$ by likelihood, i.e., to distinguish the equations $\delta_{\min} = \delta_0$ when the set of vector $x \in V_k$, in turn, is a priori ordered by the possibility of $x = x_0$ on the equation (26).

Definition 31. The problem of determining all the vectors $x \in V_k$ in equation (26), the possible error vectors of which $\delta = z_0 \overline{\oplus} xA \in \Delta_{\min}(z_0)$, we shall call the *solution* ofa (x_0, A, δ_0)-system.

It follows from definition 31 and from what has just been said, that a (x_0, A, δ_0)-system can be solved in two ways: either by setting the vectors $x \in V_k$ and then checking the condition $z_0 \overline{\oplus} xA \in \Delta_{\min}(z_0)$, or, the other way round, by trying the vectors $\delta \in V_n$, ordered by minimum of Hamming weight, until the condition $\delta \oplus z_0 = xA$ is fulfilled for some vector $x \in V_k$. It is obvious that if the set of vector V_k is not a priori ordered in the way described earlier, then the computational complexity of the first method, according to definition 31, is $O(2^k)$. For the second method, the computational complexity of the procedure is estimated by $O\left(\sum_{i=0}^{\rho_n}\binom{n}{i}\right)$, where the parameter $\rho_n = \chi(\delta_0)$ and its value is fully defined by the a priori formation characteristics of possible error vectors from the set $\Delta(z_0)$ for a particular (x_0, A, δ_0)-system.

Having said that, these two methods can be joined into one procedure. Suppose the matrix $A = \|A_k, A_{n-k}\|$, where A_k is $k \times k$ matrix and rang $A_k = k$. Respectively, the vectors $z_0 = (z_0^{(k)}, z_0^{(n-k)})$ and $\delta_0 = (\delta_0^{(k)}, \delta_0^{(n-k)})$. It is easy to show that in these terms, a (x_0, A, δ_0)-system comes down to a $(\delta_0^{(k)}, A_k^{-1}A_{n-k}, \delta_0^{(n-k)})$-system, and the equation (26) - to the equation

$$z_0^{(k)} A_k^{-1} A_{n-k} \overline{\oplus} z_0^{(n-k)} = \delta_0^{(k)} A_k^{-1} A_{n-k} \overline{\oplus} \delta_0^{(n-k)} \tag{27}$$

Let us highlight that the solution of this system, according to definition 31, is a part of the error vector $\delta_0^{(k)}$ and the vector of unknowns $x_0 \in V_k$ is determined by the equation $z_0^{(k)} \overline{\oplus} \delta_0^{(k)} = x_0 A_k$.

It is obvious that the computational complexity of such a system is estimated by $O\left(\sum_{i=0}^{\rho_k}\binom{k}{i}\right)$, where $0 \le \rho_k \le k$, and its value can depend exponentially only on parameter k. Moreover, the order of vectors of unknowns $\delta_0^{(k)} \in V_k$ by Hamming weights in this system complies with the a priori order by the probability of likelihood, which was mentioned earlier.

Suppose $H_k = \left\| \begin{matrix} A_k^{-1} A_{n-k} \\ E_{n-k} \end{matrix} \right\|$ is a $n \times (n-k)$ matrix, where E_{n-k} is an identity matrix. Then, $yH_k = e_0^{(n-k)}$ for any vector $y = xA, x \in V_k$. The vector

$z_0 H_k = \delta_0 H_k = (\varsigma_1^{(0)},\ldots,\varsigma_{n-k}^{(0)}) \in V_{n-k}$ is called *a syndrome* for the error vector $\delta_0 \in V_n$ in the (x_0, A, δ_0)-system. Obviously, in these terms $\Delta(z_0) = \{\delta \in V_n \,|\, \delta H_k = (\varsigma_1^{(0)},\ldots,\varsigma_{n-k}^{(0)})\}$.

Considering what has just been said, (x_0, A, δ_0)-systems can be divided into two groups by their methods of solution. The first group is concerned with the situation where the matrix A fixed, i.e., it is the same for different (x_0, A, δ_0)-systems, and it is possible to build the sets of error vectors $\Delta(z_0) = \{\delta \in V_n \,|\, \delta H_k = z_0 H_k\}$ in advance for all vectors $z_0 \in V_n$, order them by Hamming weights and then solve the $(\delta_0^{(k)}, A_k^{-1} A_{n-k}, \delta_0^{(n-k)})$-systems for vectors $\delta \in \Delta(z_0)$. This case is characteristic of the decoding of linear codes. If A is known, but a priori any $k \times n$ matrix of rank k for different (x_0, A, δ_0)-systems, then such a problem is referred to as that of a solution for linear equations with a skewed right-hand side. For the systems of this group the situation is possible where $n \Box k$ and the value of parameter k is such that the estimate $O(2^k)$ is not acceptable in practical terms. The problem of bringing down the maximum computational complexity, estimated by $O(2^k)$, becomes central in finding the solution of such (x_0, A, δ_0)-systems.

Further on we will look at (x_0, A, δ_0)-systems as G-factorization of binary matroids.

Suppose $S = \{1, 2, \ldots, n\}$ and $M \in \mathfrak{N}(S)$ is a binary matroid of rank $r_M(S) = k$. For any base $B \in \mathfrak{B}_M$ of the matroid M and the respective base $S - B = B^* \in \mathfrak{B}_{M^*}$ of the dual matroid M^* the families of fundamental cycles would be of the forms $\mathfrak{C}_M(B) = \{C(b^*, B) \in \mathfrak{R}_M \,|\, b^* \in B^*\}$ and $\mathfrak{C}_{M^*}(B^*) = \{C^*(b, B^*) \in \mathfrak{R}_{M^*} \,|\, b \in B\}$, respectively.

Suppose that the binary matroid $H^*(D_0^*) \in \mathfrak{N}(S)$ is a G-factor of the matroid M^*, generated by a modular filter $\Phi_{M^*}^* = \Phi(\mathbb{R}_{M^*}(D_0^*))$, where the subset $D_0^* \in \mathfrak{F}_{M^*}, D_0^* \neq \varnothing$. For the respective semi-matroid $\mathbb{R}(M^*, D_0^*)$ and any cycles $D^* \in \mathbb{R}_{M^*}(D_0^*), D^* \neq D_0^*$, by definition, $D^* = D_0^* \oplus \sum C^*$, where cycles $C^* \in \mathfrak{R}_{M^*}$ and $D_0^* \cap C_0^* \neq \varnothing$. From this, considering (1), we get that $D^* = D_0^* \oplus \sum_{b \in J} \oplus C^*(b, B^*)$ for some subset $J \subseteq B$.

For brevity's sake, we shall call the combination of inter-connected objects above a (J, M, D^*)*-scheme.*

Let us fix a base $B_0 \in \mathfrak{B}_M$ and (respectively) a base $S - B_0 = B_0^* \in \mathfrak{B}_{M^*}$. Suppose that the cycle $C_0^* \in \mathbb{R}_{M^*}(D_0^*)$ is set. Then $C_0^* = D_0^* \oplus \sum_{b \in J_0} \oplus C^*(b, B_0^*)$, and the subset $J_0 \subseteq B_0$ is defined unequivocally. Let us remind ourselves that, according to theorem 12, $\mathbb{R}_{M^*}(D_0^*) = \mathbb{R}_{M^*}(C_0^*)$.

Let us denote

$$\mathfrak{D}(C_0^*) = \{D^* = C_0^* \oplus \sum_{b \in J} \oplus C^*(b, B_0^*) | J \subseteq B_0\}$$
.

It is obvious that $\mathbb{R}_{M^*}(D_0^*) \subseteq \mathfrak{D}(C_0^*)$. If the set $D^* \in \mathfrak{D}(C_0^*)$ and $D^* \notin \mathbb{R}_{M^*}(D_0^*)$, then there will be found a cycle $C^* \in \mathfrak{R}_{M^*}$, such that $C^* \subset D^*$ and $C^* \cap D_0^* = \varnothing$. Indeed, as $D^* = D_0^* \oplus \sum_{b \in J \oplus J_0} \oplus C^*(b, B_0^*)$, then, by the axiom of cycles for binary matroid, $D^* = D_0^* \oplus \sum C^*$, $C^* \in \mathfrak{R}_{M^*}$, and if $C^* \cap D_0^* \neq \varnothing$ for all cycles in the latter sum, then $D^* \in \mathbb{R}_{M^*}(D_0^*)$.

Thus, any set $D^* \in \mathfrak{D}(C_0^*)$, for which $D^* \notin \mathbb{R}_{M^*}(D_0^*)$ is of the form $D^* = \overline{D}^* + \sum \overline{C}^*$, where $\overline{C}^* \in \mathfrak{R}_{M^*}$ and $\overline{D}^* \in \mathbb{R}_{M^*}(D_0^*)$. Summing up, we get $\mathfrak{D}(C_0^*) = \{D^* = D_0^* \oplus \sum_{b \in J \oplus J_0} \oplus C^*(b, B_0^*) | J \subseteq B_0\} =$

$\{D^* = D_0^* \oplus \sum_{b \in J} \oplus C^*(b, B_0^*) | J \subseteq B_0\}$, and the set $\sum_{b \in J} \oplus C^*(b, B_0^*)$ for some subset $J \subseteq B_0$ can contain cycles $\overline{C}^* \in \mathfrak{R}_{M^*}$, for which $\overline{C}^* \cap D_0^* = \varnothing$.

The form of cycle $C_0^* \in \mathbb{R}_{M^*}(D_0^*)$ implies that any elements $b \in B_0$ fulfill the following conditions:

1. $b \in C_0^*$. If $b \in J_0$, then $b \notin D_0^*$, and, conversely, if $b \notin J_0$, then $b \in D_0^*$;
2. $b \notin C_0^*$. In this case either $b \in J_0$ and $b \in D_0^*$, or $b \notin J_0$ and $b \notin D_0^*$.

The similar situation is for any set $D^* \in \mathfrak{D}(C_0^*)$ and the correspondent subset $J \subseteq B_0$. Particularly, for the cycle $\overline{C}^* \in \mathfrak{R}_{M^*}$ such that $\overline{C}^* \subset D^*$ and $\overline{C}^* \cap D_0^* = \varnothing$, if $b \in \overline{C}^*$, then the element $b \notin D_0^*$, and hence $b \in J$.

Let $\mathfrak{D}_{\min}(C_0^*) \subseteq \mathfrak{D}(C_0^*)$ and the subset $D_{\min}^* \in \mathfrak{D}(C_0^*)$ belongs to the set $\mathfrak{D}_{\min}(C_0^*)$ if and only if $\left|D_{\min}^*\right| = \min_{D^* \in \mathfrak{D}(C_0^*)} \left|D^*\right|$.

Definition 32. For a given $C_0^* \in \mathbb{R}_{M^*}(D_0^*)$ the problem of finding all sets $D^* \in \mathfrak{D}(C_0^*)$, for which $D^* \in \mathfrak{D}_{\min}(C_0^*)$, we shall call the *inversion of the* (J_0, M, D_0^*)*-scheme,* and the set $\mathfrak{D}_{\min}(C_0^*)$ – the *result of the inversion of the* (J_0, M, D_0^*) *- scheme.*

According to the abovementioned, $\mathfrak{D}_{\min}(C_0^*) \subseteq \mathbb{R}_{M^*}(C_0^*) = \mathbb{R}_{M^*}(D_0^*)$, and from definition 32 it follows that the problem of the inversion of the

(J_0, M, D_0^*) -scheme comes down to the problem of building all minimal by cardinality cycles of the semi-matroid $\mathbb{R}(M^*, D_0^*)$.

If $M(A) \in \mathfrak{N}(S)$ is a binary matroid, isomorphic to a vector matroid, built on $k \times n$ a matrix A of rank k over the field $GF(2)$, $J_0(x)$ and $D_0^*(\delta)$ are the subsets of base B_0 and the set S, respectively, generated by non-zero elements of vectors $x \in V_k$ and $\delta \in V_n$, then the (x_0, A, δ_0) -system can be represented as a $(J_0(x_0), M(A), D_0^*(\delta_0))$ -scheme. Therefore, any characteristic of the parameters of a (x_0, A, δ_0) -system or the method of its solution has a matroid interpretation.

For example, suppose that $|C^*| \geq d+1$ for any cycle $C^* \in \mathfrak{R}_{M^*(A)}$. If $D^* \in \mathbb{R}_{M^*(A)}(D_0^*(\delta_0))$, then $D^* = D_0^*(\delta_0) \oplus \sum C^*$, where $C^* \in \mathfrak{R}_{M^*(A)}$ and $C^* \cap D_0^*(\delta_0) \neq \varnothing$. Therefore, $|D^*| = |D_0^*(\delta_0)| + |\sum C^*| - 2|D_0^*(\delta_0) \cap \sum C^*| \geq |\sum C^*| - |D_0^*(\delta_0)| \geq \frac{d}{2} + 1$, as soon as the inequality $|D_0^*(\delta_0)| \leq \frac{d}{2}$ is fulfilled. If $D^* \notin \mathbb{R}_{M^*(A)}(D_0^*(\delta_0))$, then $D^* = \overline{D}^* + \sum \overline{C}^*$ and $|D^*| > |\overline{D}^*|$. As a result, in any case we get that $|\mathfrak{D}_{\min}(C_0^*)| = 1$ and $D_{\min}^* = D_0^*(\delta_0)$, if $|D_0^*(\delta_0)| \leq \frac{d}{2}$, which matches the analogous condition for a (x_0, A, δ_0) -system.

As with the solution of a (x_0, A, δ_0) -system, there are two methods for the inversions of a (J_0, M, D_0^*) -system. At that, the conditions outlined above for elements $b \in B_0$ imply that for any set $D^* \in \mathfrak{D}(C_0^*)$ the respective set $J \subseteq B_0$ can be constructed based on the known set $D^* \cap B_0$ alone. It is easy to show that this fact corresponds with the shift from a (x_0, A, δ_0) -system to a $(\delta_0^{(k)}, A_k^{-1} A_{n-k}, \delta_0^{(n-k)})$ -system.

Suppose $b_1 \in B_0$ and G-mapping $M \xrightarrow{\Phi} M_1$ is generated by a modular filter $\Phi_M = \Phi(\mathbb{R}_M(b_1))$. Then $b_1 \in \Im_{M_1}(\varnothing)$ is a loop in the matroid M_1 and does not belong to the bases of the matroid M_1. Respectively, $\mathbb{R}_M(b_1) = \{b_1\} + \{C - b_1 | C \in \Re_M, b_1 \in C\}$ and $\Re_M(b_1) = \{C \in \Re_M | b_1 \notin C\} + \Im_M(\varnothing)$. Let us remind ourselves that the set of loops $\Im_M(\varnothing)$ belongs to any G-factor of the matroid M; in particular, $\Im_M(\varnothing) \subset \Im_{M_1}(\varnothing)$. Further, without any loss of generality, we shall assume that $\Im_M(\varnothing) = \varnothing$.

Consider the conditions above for an element $b_1 \in B_0$.

1. $b_1 \in C_0^*$. For the case of $b_1 \in J_0$ and $b_1 \notin D_0^*$ let us denote $J_1 = J_0 - b_1$, $C_1^* = C_0^* \oplus C^*(b_1, B_0^*)$ and $D_1^* = D_0^*$. If $b_1 \notin J_0$ and $b_1 \in D_0^*$, then $J_1 = J_0$, $C_1^* = C_0^* \oplus b_1 = C_0^* - b_1$ and $D_1^* = D_0^* - b_1$.
2. $b_1 \notin C_0^*$. In this case, if $b_1 \in J_0$ and $b_1 \in D_0^*$, then $J_1 = J_0 - b_1$, $C_1^* = C_0^* \oplus C^*(b_1, B_0^*)$ and $D_1^* = D_0^* - b_1$. If $b_1 \notin J_0$ and $b_1 \notin D_0^*$, then $J_1 = J_0$, $C_1^* = C_0^*$ and $D_1^* = D_0^*$.

It easy to show that in these terms the problem of the inversion of a (J_0, M, D_0^*)-system comes down to the problem of the inversion of a (J_1, M_1, D_1^*)-system where the couples (J_1, D_1^*) fulfill the conditions 1 and 2. Let us point out that there are two such couples in both cases.

Consider the G-factorization:

$$M = M_0 \xrightarrow{\Phi_1} M_1 \xrightarrow{\Phi_2} \ldots \xrightarrow{\Phi_{r-1}} M_{r-1} \xrightarrow{\Phi_r} M_r, \tag{28}$$

where the modular filters $\Phi_i = \Phi(\mathbb{R}_{M_{i-1}}(b_i)), i = \overline{1,r}$, $\{b_1,...,b_r\} \subset B_0$. By means of iterations it can be demonstrated that the equations

$\mathbb{R}_{M_{r-1}}(b_r) = \{b_r\} + \{C - b_r \,|\, C \in \mathfrak{R}_{M_{r-1}}, b_r \in C\}$ and

$\mathfrak{R}_{M_{r-1}}(b_r) = \{b_1\} + \{b_2\} + ... + \{b_{r-1}\} + \{C \in \mathfrak{R}_{M_{r-1}} \,|\, b_r \notin C\}$ hold true.

It is also clear that for each $i = \overline{1,r}$ there exist two couples of subset (J_i, D_i^*), that fulfill the conditions 1 and 2 on each step of the G-factorization (28), and, as a result, there would be 2^r different couples of subsets (J_r, D_r^*) and, respectively, 2^r different (J_r, M_r, D_r^*)-schemes.

Let us show that for any of these options the weight characteristics can be constructed and, thus, they can be ordered by likelihood.

For any element $b \in B_0$ let us denote the set of cycles of the length $l, 2 \le l \le r_M(S) + 1$, that contain the element b, by $\mathfrak{C}_b^{(l)}(M) = \{C \in \mathfrak{R}_M \,|\, b \in C, |C| = l\}$. Let us note that the element b cannot be a loop and therefore $l \ge 2$. For any element $a \in S$ suppose

$$\gamma_a(C_0^*) = \begin{cases} -1, \text{ if } a \in C_0^*, \\ 1, \text{ if } a \notin C_0^*. \end{cases}$$

Suppose also that $D_0^* = \varnothing$ and, respectively, $C_0^* = \sum_{b \in J_0} \oplus C^*(b, B_0^*)$. In these terms, for any element $b \in B_0$ let us define the numerical functions

$$W_b^{(0)}(C_0^*) = \sum_{(b,a_1,a_2) \in \mathfrak{C}_b^{(3)}(M)} \left|\gamma_{a_1}(C_0^*) + \gamma_{a_2}(C_0^*)\right| + \sum_{(b,a) \in \mathfrak{C}_b^{(2)}(M)} \gamma_a(C_0^*),$$

$$W_b^{(1)}(C_0^*) = \sum_{(b,a_1,a_2) \in \mathfrak{C}_b^{(3)}(M)} \left|\gamma_{a_1}(C_0^*) - \gamma_{a_2}(C_0^*)\right| - \sum_{(b,a) \in \mathfrak{C}_b^{(2)}(M)} \gamma_a(C_0^*). \quad (29)$$

Because for binary matroids $|C \cap C^*| = 2t, t \ge 0$, for any cycles $C \in \mathfrak{R}_M, C^* \in \mathfrak{R}_{M^*}$ and $C_0^* = \sum_{b \in J_0} \oplus C^*(b, B_0^*) = \sum C^*$, $C^* \in \mathfrak{R}_{M^*}$, so that for any cycle $C \in \mathfrak{C}_b^{(2)}(M) + \mathfrak{C}_b^{(3)}(M)$ either $\left|C \cap C^*\right| = 0$ for all the cycles C^*, or there will be found a single cycle C^*, for which $\left|C \cap C^*\right| = 2$. Therefore, $W_b^{(0)}(C_0^*) = 2\left|\mathfrak{C}_b^{(3)}(M)\right| + \left|\mathfrak{C}_b^{(2)}(M)\right|$ and $W_b^{(1)}(C_0^*) = -\left|\mathfrak{C}_b^{(2)}(M)\right|$, if $b \notin C_0^*$, which means that $b \notin J_0$. Otherwise, $W_b^{(0)}(C_0^*) = -\left|\mathfrak{C}_b^{(2)}(M)\right|$ and $W_b^{(1)}(C_0^*) = 2\left|\mathfrak{C}_b^{(3)}(M)\right| + \left|\mathfrak{C}_b^{(2)}(M)\right|$, if $b \in C_0^*$ and $b \in J_0$.

In general cases, when $D_0^* \neq \varnothing$, the weight functions $W_b^{(0)}(C_0^*)$ and $W_b^{(1)}(C_0^*)$ may not reach their extreme values. For example, suppose $b_0 \notin C_0^*$ and $b_0 \notin J_0$. This means that $b_0 \notin D_0^*$ and $\left(\mathfrak{C}_{b_0}^{(2)}(M) + \mathfrak{C}_{b_0}^{(3)}(M)\right) \cap \left(\sum_{b \in J_0} \oplus C(b, B_0^*) \right) = \varnothing$. However, the situation of $a \in D_0^* - C_0^*$ or $a_i \in D_0^* - C_0^*, i = 1,2$, for the cycles $(b_0, a) \in \mathfrak{C}_b^{(2)}(M)$ and $(b_0, a_1, a_2) \in \mathfrak{C}_b^{(3)}(M)$ may occur. In this case, the weight functions (29) would be of the form

$$W_{b_0}^{(0)}(C_0^*) = 2\left[\left|\mathfrak{C}_{b_0}^{(3)}(M)\right| - \left|\left\{(b_0, a_1, a_2) \in \mathfrak{C}_{b_0}^{(3)}(M) \middle| \left|(a_1, a_2) \cap (D_0^* - C_0^*)\right| = 1\right\}\right|\right] +$$
$$+\left|\left\{(b_0, a) \in \mathfrak{C}_{b_0}^{(2)}(M) \middle| a \notin (D_0^* - C_0^*)\right\}\right| - \left|\left\{(b_0, a) \in \mathfrak{C}_{b_0}^{(2)}(M) \middle| a \in (D_0^* - C_0^*)\right\}\right|$$
$$W_{b_0}^{(1)}(C_0^*) = 2\left|\left\{(b_0, a_1, a_2) \in \mathfrak{C}_{b_0}^{(3)}(M) \middle| \left|(a_1, a_2) \cap (D_0^* - C_0^*)\right| = 1\right\}\right| +$$
$$+\left|\left\{(b_0, a) \in \mathfrak{C}_{b_0}^{(2)}(M) \middle| a \in (D_0^* - C_0^*)\right\}\right| - \left|\left\{(b_0, a) \in \mathfrak{C}_{b_0}^{(2)}(M) \middle| a \notin (D_0^* - C_0^*)\right\}\right|. \tag{30}$$

A similar situation occurs also for alternative conditions for elements $b \in B_0$. Obviously, in the general case, when $D_0^* \neq \varnothing$, the efficiency of the separation of hypotheses $b \in D_0^*$ or $b \notin D_0^*$ by means of weight functions (30) depends entirely on the ratio of values $\left|\mathfrak{C}_b^{(2)}(M)\right|$, $\left|\mathfrak{C}_b^{(3)}(M)\right|$ and $\left|D_0^*\right|$. In this connection let us note that on each step of the G-factorization (28) the number of cycles in the sets $\mathfrak{C}_b^{(2)}(M)$ and $\mathfrak{C}_b^{(3)}(M)$ can only increase. For example, if $r = \text{int}(\log n)$, i.e., $n \cong 2^r$, and the number of cycles in the set $\mathfrak{C}_b^{(3)}(M)$ can be close to its maximum value of $n/2$. Summarizing what has just been said, we get that on each step $r, 1 \le r \le k$, of the G-factorization (28) by means of the weight functions (29) the alternative couples (J_r, D_r^*) can be ordered for the respective (J_r, M_r, D_r^*)-schemes, so that those with the most likelihood are left for the next step, thus decreasing the maximal possible computational complexity, estimated by $O(2^k)$.

Let us point out that it makes sense to use this procedure for the solution of the systems of linear equations with skewed right-hand side, for which the estimate $O(2^k)$ is not acceptable from the practical point of view.

Let us now consider the situation identical to the decoding of linear (n,k)-codes.

Suppose $B_0^* = \{b_1^*, b_2^*, \ldots, b_{n-k}^*\}$ and $D^* \in \mathfrak{D}(C_0^*)$.

Let us define an integer vector

$$W_M(D^*, B_0) = \left(\left|C_M(b_1^*, B_0) \cap D^*\right|, \ldots, \left|C_M(b_{n-k}^*, B_0) \cap D^*\right|\right). \tag{31}$$

If $D^* \in \mathbb{R}_{M^*}(D_0^*)$, then theorems 22 and 24 imply that $\left|C_M(b_i^*, B_0) \cap D^*\right| = 2t_i + \varsigma_i^{(0)}$, where $t_i \ge 0, \varsigma_i^{(0)} = 0$, if $b_i^* \in D_0^*$, and $\varsigma_i^{(0)} = 1$, if $b_i^* \in B_0^* - D_0^*$, for all $i = \overline{1, n-k}$. If $D^* = \overline{D}^* + \sum \overline{C}^*$, where $\overline{D}^* \in \mathbb{R}_{M^*}(D_0^*)$ and $\overline{C}^* \in \mathfrak{R}_{M^*}$, then $\left|\left(\sum \overline{C}^*\right) \cap C_M(b_i^*, B_0)\right| = 2t$, $t \ge 0$, for any $i = \overline{1, n-k}$. Thus, we arrive at the fact that for all the sets $D^* \in \mathfrak{D}(C_0^*)$ the vector $W_M(D^*, B_0)$ is of the form: $W_M(D^*, B_0) = \left(2t_1 + \varsigma_1^{(0)}, \ldots, 2t_{n-k} + \varsigma_{n-k}^{(0)}\right)$.

If we now assume that we are considering a $(J_0(x_0), M(A), D_0^*(\delta_0))$-scheme, then the vector $\left(\varsigma_1^{(0)}, \ldots, \varsigma_{n-k}^{(0)}\right)$ is none other than a syndrome in the (x_0, A, δ_0)-system for the error vector δ_0. In turn, the set of cycles

$\mathbb{R}_{M^*(A)}(D_0^*(\delta_0))$ is matched by the set $\Delta(z_0)$ of all possible error vectors with the syndrome $\left(\varsigma_1^{(0)},\ldots,\varsigma_{n-k}^{(0)}\right)$, with the exceptions that there are no sums of non-intersection cycles $\mathbb{R}_{M^*(A)}(D_0^*(\delta_0))$ and cycles $\mathfrak{R}_{M^*(A)}$. However, as we understand the sets $D^* \in \mathfrak{D}_{\min}(C_0^*(\delta_0))$, where $C_0^*(\delta_0) \in \mathbb{R}_{M^*(A)}(D_0^*(\delta_0))$, as the result of the inversion of a respective scheme, this difference is not crucial.

Let us demonstrate that the problem of the inversion of a (J,M,D_0^*)-scheme, in turn, comes down to the problem of building families of cycles of G-lifts of matroid M, the solution algorithm for which has been offered in part III.

As has already been noted, the inversion of a (J_0,M,D_0^*)-scheme means finding the minimal by cardinality cycles of the semi-matroid $\mathbb{R}(M^*,D_0^*)$ in the diagram

$$\begin{array}{ccc} H(D_0^*) & \xrightarrow{\;D_0\;} & M \\ \updownarrow & & \updownarrow \\ H^*(D_0^*) & \xleftarrow{\;D_0^*\;} & M^* . \end{array}$$

The matroid $H(D_0^*)$ is a G-lift of the matroid M by definition, and the matroid $M = M(D_0)$ itself is a G-factor of the matroid $H(D_0^*)$ for some set D_0. For the given cycle $C_0^* \in \mathbb{R}_{M^*}(D_0^*)$, base $B_0^* \in \mathfrak{B}_{M^*}$ such that $C_0^* \subseteq B_0^*$, and an arbitrary element $d_0^* \in C_0^*$ the family of fundamental cycles $\mathfrak{C}_{H(C_0^*)}\left(B_0 \cup d_0^*\right)$, where the base $B_0 = S - B_0^* \in \mathfrak{B}_M$, can be built, using theorem 24. It is obvious that $\mathfrak{R}_{H(C_0^*)} = \left\{C = \sum \oplus D \middle| D \in \mathfrak{C}_{H(C_0^*)}\left(B_0 \cup d_0^*\right) \text{ and } C - \min\right\}$ is the family if cycles of the matroid $H(C_0^*)$. It has already been mentioned that the fundamental cycle $C_M\left(d_0^*, B_0\right)$ of the matroid M, in compliance with

(23), would be a cycle of the semi-matroid $\mathbb{R}\left(H(C_0^*),M\right)$, treated as a pseudo-matroid. Therefore, the matroid $M=M(D_0)$ is a G-factor of the matroid $H(C_0^*)$ for any set $D_0\in\mathbb{R}_{H(C_0^*)}(C_M(d_0^*,B_0))$. According to theorem 22, for any cycles $D\in\mathbb{R}_{H(C_0^*)}(C_M(d_0^*,B_0))$ and $D^*\in\mathbb{R}_{M^*}(C_0^*)$ the equation $\left|D\cap D^*\right|=2\mu+1,\mu\geq 0$ must hold true. Thus, any set D^*, for which $\left|D\cap D^*\right|=2\mu+1,\mu\geq 0,$ for all cycles $D\in\mathbb{R}_{H(C_0^*)}(C_M(d_0^*,B_0))$, belongs to the family of cycles $\mathbb{R}_{M^*}(C_0^*)$ of the semi-matroid $\mathbb{R}\left(M^*,C_0^*\right)$, and if its cardinality $\left|D^*\right|$ is minimal, then $D^*\in\mathfrak{D}_{\min}(C_0^*)$, i.e., considering $\mathbb{R}_{M^*}(C_0^*)=\mathbb{R}_{M^*}(D_0^*)$, D^* is the result of the inversion of the (J_0,M,D_0^*)-scheme.

So, we obtain the following algorithm for the solution of the problem of inversion:

1. For a given cycle $C_0^*\in\mathbb{R}_{M^*}(D_0^*)$ a base $B_0^*\in\mathfrak{B}_{M^*}$ of the dual matroid M^* is chosen, such that $C_0^*\subseteq B_0^*$.
2. A fundamental cycle $C_M\left(d_0^*,B_0\right)$ of the matroid M for the base $B_0=S-B_0^*\in\mathfrak{B}_M$ and an arbitrary element $d_0^*\in C_0^*$ is found.
3. According to theorem 24, the family of fundamental cycles of the matroid $H(C_0^*)$ is built for the base $B_0\cup d_0^*$:
$$\mathfrak{C}_{H(C_0^*)}\left(B_0\cup d_0^*\right)=$$
$$=\left\{C_M\left(b^*,B_0\right)\middle|b^*\in B_0^*-C_0^*\right\}+\left\{C_M\left(b^*,B_0\right)\oplus C_M\left(d_0^*,B_0\right)\middle|b^*\in C_0^*-d_0^*\right\}.$$
4. Based on the family $\mathfrak{C}_{H(C_0^*)}\left(B_0\cup d_0^*\right)$, the family of cycles of the matroid $H(C_0^*)$ is built:
$$\mathfrak{R}_{H(C_0^*)}=\left\{C=\sum\oplus D\middle|D\in\mathfrak{C}_{H(C_0^*)}\left(B_0\cup d_0^*\right)\text{ and }C-\min\right\}.$$

5. We get the set of cycles of the semi-matroid $\mathbb{R}\left(H(C_0^*), C_M\left(d_0^*, B_0\right)\right)$, according to the equation:
$$\mathbb{R}_{H(C_0^*)}\left(C_M\left(d_0^*, B_0\right)\right) = \mathfrak{R}_M - \mathfrak{R}_{H(C_0^*)}.$$
6. The set $\mathfrak{D}(C_0^*)$ is of the form $\mathfrak{D}(C_0^*) = \left\{D_0^* \middle| \left|D_0 \cap D_0^*\right| = 2\mu + 1, \mu \geq 0,\right.$ for all $D_0 \in \mathbb{R}_{H(C_0^*)}\left(C_M\left(d_0^*, B_0\right)\right)$ and D_0^* – min$\}$.
7. The resulting set would be $\mathfrak{D}_{\min}(C_0^*) \subseteq \mathfrak{D}(C_0^*)$.

Let us illustrate this with an example.

Example 6. Suppose $S = \{1,2,3,4,5,6,7\}$ and the matroid M is the Fano matroid, represented by the matrix from example 1.

It easy to show that the family of cycles of the dual matroid $\mathfrak{R}_{M^*} = \{\{1,2,3,7\},\{1,2,5,6\},\{1,3,4,6\},\{1,4,5,7\},\{2,3,4,5\},\{2,4,6,7\}, \{3,4,5,6\}\}$ and, respectively, $\mathfrak{R}_M = \mathfrak{R}_{M^*} + \{\{1,2,4\},\{1,3,5\},\{1,6,7\},\{2,3,6\},\{2,5,7\},\{3,4,7\},\{4,5,6\}\}$. Let us suppose that $D_0^* = \{2,4\}$ and the cycle $C_0^* = \{2,4,6,7\} \oplus D_0^* = \{6,7\}$ is given. Obviously, $\mathbb{R}_{M^*}(C_0^*) = \mathbb{R}_{M^*}(D_0^*) = \{\{2,4\},\{3,5\},\{6,7\},\{1,3,4,7\},$ $\{1,4,5,6\},\{1,2,5,7\},\{1,2,3,6\},\{2,3,5,6\}\}$. Let us note that the minimal by cardinality cycles of the semi-matroid $\mathbb{R}(M^*, D_0^*)$ would be the sets $\{2,4\},\{3,5\}$ and $\{6,7\}$.

Suppose the element $d_0^* = \{6\} \in C_0^*$ and let us choose the base $B_0^* = \{4,5,6,7\}$, so that $C_0^* \subseteq B_0^*$ and $B_0 = \{1,2,3\} \in \mathfrak{B}_M$. Then $C_M(d_0^*, B_0) = \{2,3,6\}$ is a fundamental cycle for the element d_0^* in the base B_0. The family of fundamental cycles of the matroid $H(C_0^*)$ for the base $B_0 \cup d_0^*$, according to theorem 24, is of the form $\mathfrak{C}_{H(C_0^*)}\left(B_0 \cup d_0^*\right) = \{\{1,2,4\},\{1,3,5\},\{1,6,7\}\}$. Respectively, $\mathfrak{R}_{H(C_0^*)} = \{\{1,2,4\},\{1,3,5\},\{1,6,7\},\{2,3,4,5\},\{2,4,6,7\},\{3,5,6,7\}\}$. Thus, $\mathbb{R}_{H(C_0^*)}\left(C_M(d_0^*, B_0)\right) = \mathfrak{R}_M - \mathfrak{R}_{H(C_0^*)} = \{\{2,3,6\}, \{2,5,7\}, \{3,4,7\},\{4,5,6\},$

$\{1,2,3,7\},\{1,2,5,6\},\{1,3,4,6\},\{1,4,5,7\}\}$. It easy to see that, according the algorithm presented above, $\mathfrak{D}_{\min}(C_0^*)=\{\{2,4\},\{3,5\},\{6,7\}\}$.

To conclude, we can infer that the classical problem of the solution of a system of linear equations with a skewed right-hand side, or the decoding problem for linear codes, allows a matroid interpretation not only in terms of its statement, but also in terms of the realization of its solution methods. In particular, the general theoretical outcomes for semi-matroids, obtained in the second part of this study, have a purely practical application for this case.

3. *G*-CODES

In this section a new coding method (*G*-coding) is described, which generates a consequence of code words, blocked into a Markov chain. It is demonstrated that the traditional convolutional linear codes are representable as *G*-codes.

Consider a linear (n,k)-code $Y(A)=\{xA \mid x\in V_k\}$. It is known [14] that there exist $\prod_{i=0}^{k-1}(2^k-2^i)$ different matrixes A, which generate this linear code $Y(A)$. The vectors' matroids, built based on these matrixes, are evidently isomorphic. The vectors $y\in Y(A)$ are called *code words*, and the equation $y=xA$ – the *coding equation* for the vectors $x\in V_k$. When coding a vector sequence, a situation may occur where vectors $x_i\in V_k$, and therefore the respective code words $y_i\in V_n$, are dependent for all $i\geq 0$.

For example, a convolutional (r,s,l)-code is generated by a $s(l+1)\times r$ matrix $A=\begin{Vmatrix}A_0\\ \dots\\ A_l\end{Vmatrix}$, and the coding equation is of the form $w_i=(v_i,v_{i-1},\dots,v_{i-l})A=v_iA_0\overline{\oplus}v_{i-1}A_1\overline{\oplus}\dots\overline{\oplus}v_{i-l}A_l$, where $v_i\in V_s$ and $w_i\in V_r$.

Further, let us consider a coding procedure that generates a sequence of code words, blocked into a Markov chain, where the convolutional coding would be a special case.

Suppose $k \times n$ matrixes A and B of rank k and a $m \times n$ matrix $C12$ of rank m, $m \geq 1$ are set over the field of $GF(2)$.

Definition 33. We shall call the procedure for building the sequence of code words

$$y_i = x_{i-1}A \overline{\oplus} u_i C \overline{\oplus} x_i B, i \geq 1, \tag{32}$$

the G *-coding* for the sequence of vectors $\{(u_i, x_i) \in V_{k+m} | i \geq 0\}$, where $u_i \in V_m$ and $x_i \in V_k$, and the set of matrixes $(A, C, B)-$ the *generating matrixes of* G *-code* with parameters n, k and m.

It important to highlight the crucial importance of the condition $m \geq 1$, from which it follows that, in the sequence $(u_i, x_i), i \geq 0$ which is being coded, there would always be elements that take part in building only one code word $y_i, i \geq 1$. It is also clear that the G -code rate is $R_G = (k+m)/n$ and $R_G > k/n$.

Let us demonstrate that classical convolutional linear codes are not G -codes. Indeed, for any vector sequence $\{v_i \in V_s | i \geq 0\}$ that is coded by a convolutional (r, s, l)-code, the sequence of code words $\{w_i \in V_r | i \geq l\}$ is built according to the equation

$$w_i = v_i A_0 \overline{\oplus} v_{i-1} A_1 \overline{\oplus} \ldots \overline{\oplus} v_{i-l} A_l . \tag{33}$$

If $l = 1$, then $w_i = v_i A_0 \overline{\oplus} v_{i-1} A_1$, and we arrive at a contradiction with the condition $m \geq 1$. If $l \geq 2$, then the coded vector sequence $v_i, i \geq 0$, there are no elements that take part in building only one code word $w_i, i \geq l$. In particular, the convolutional $(r,s,2)$-code is generated by a set of matrixes (A_0,A_1,A_2), but is not G-code.

Thus, the code word sequence of the convolutional code (33) cannot be a code word sequence of the G-code (32), and vice versa. At the same time, we will further demonstrate that any convolutional code can be presented as a G-code.

Statement 29. *An arbitrary convolutional* (r,s,l)*- code is presentable as a* G*-code with parameters* $n=r(l+1+t), k=s(l+t)$ *and* $m=s$, *where* $t\geq 0$.

Proof. Suppose for a convolutional (r,s,l)-code the matrixes A_{sl}, C_s and B_{sl} are of the form:

$$A_{sl}=\left\|\begin{matrix} A_l & \varnothing\ldots & \varnothing & \varnothing \\ A_{l-1} & A_l\ldots & \varnothing & \varnothing \\ \cdot & \cdot & \cdot & \cdot \\ A_1 & A_2\ldots & A_l & \varnothing \end{matrix}\right\|,\ B_{sl}=\left\|\begin{matrix} \varnothing & A_0 & A_1\ldots & A_{l-1} \\ \varnothing & \varnothing & A_0\ldots & A_{l-2} \\ \cdot & \cdot & \cdot & \cdot \\ \varnothing & \varnothing & \varnothing\ldots & A_0 \end{matrix}\right\|,\ C_s=\|A_0A_1\ldots A_l\|, \quad (34)$$

where $\varnothing$ is a $k\times n$ zero matrix.

Then, from the equation (33) we get that

$$(w_i,w_{i+1},\ldots,w_{i+l})=(v_{i-l},\ldots,v_{i-1})A_{sl}\overline{\oplus}v_iC_s\overline{\oplus}(v_{i+1},\ldots,v_{i+l})B_{sl}. \quad (35)$$

Therefore, the sequence of code words $w_i,w_{i+1},\ldots,w_{i+l}$ on steps $i,i+1,\ldots,i+l$, obtained by convolutional (r,s,l)-coding, coincides as the sequence of bits of length $r(l+1)$ with the code word of one step of G-coding, where the G-code is generated by the matrixes (34) with parameters $n=r(l+1), k=sl$ and $m=s$. So, if vector sequence $\{v_i\in V_s\,|i\geq 0\}$, coded by a convolutional linear (r,s,l)-code, is presented as a sequence of vectors $\{(v_{(l+1)i-1},(v_{(l+1)i},\ldots,v_{(l+1)i+l-1}))\in V_{s(l+1)}\,|i\geq 0\}$, where $v_{-1}=e_0^{(s)}$, then, given the same initial conditions, the resulting code word sequences, presented as sequences of bits, would coincide for the convolutional and for the G-code, generated by the matrixes (34).

It is also easy to show that the matrixes $A_{s(l+t)}, C_s, B_{s(l+t)}$, analogous to the matrixes (34) with parameters $n = r(l+1+t)$, $k = s(l+t)$ and $m = s$, can be built so that the respective G-code would be of rate $R_G = s(l+1+t)/r(l+1+t), t \geq 0$.□

Let us note that the rate of G-code, built according to statement 29, is numerically equal to the rate of the initial code $R = s/r$.

From now on we shall call equation (32) the *"i-th step" of G-coding.*

Let us highlight that the generating matrixes and their parameters for arbitrary G-code may not coincide with the respective characteristics of the G-codes that are the representations of convolutional codes. In that sense, considering statement 29, G-coding can be considered a generalized form of convolutional coding.

Let us now describe the decoding procedure of an arbitrary G-code with parameters n, k and m, analogous to the Viterbi decoder for convolutional codes.

On the vector set V_n some weight function is defined, and on the i-*th* step of decoding, according to the results of all previous steps, in the memory of "current cumulative weights" the weights of all 2^k alternative vectors $x_{i-1} \in V_k$ are stored. For any of these 2^{2k+m} vectors $(x_{i-1}, u_i, x_i) \in V_{2k+m}$ *the* weights, *equal* to the "current cumulative weight" of vectors x_{i-1} plus the weights of the respective code words $y_i = x_{i-1}A \overline{\oplus} u_i C \overline{\oplus} x_i B$, are calculated. For every fixed vector $x_i \in V_k$, based on these weights, according to the maximum likelihood principle, one vector $(x_{i-1}, u_i) \in V_{k+m}$ is chosen. This vector itself is written into the "backward memory", while its weight is written into the "current cumulative weights" memory for the next coding step. So, as the result of an *i-th* step decoding of *a* G-code a memory of new "current cumulative weights" for all 2^k vectors $x_i \in V_k$ is formed, and for each of them in the "backward memory" the most likelihood of the alternative vectors $(x_{i-1}, u_i) \in V_{k+m}$.

Further, for the best in terms of maximum likelihood principle vector $x_i \in V_k$ by the "backward memory" and for the given parameter L a sequence of vectors $x_i, (x_{i-1}, u_i), \ldots, (x_{i-L}, u_{i-L+1})$ is built. If i is the last step of decoding, then the whole sequence is the result of decoding on the *i-th* step; *if* not, then only the vector (x_{i-L}, u_{i-L+1}) is. The initial sequence of vectors $(u_i, x_i) \in V_{k+m}, i \geq 1$ is obviously formed by the results of decoding.

It is easy to show that if the weight of the code word $(w_i, w_{i+1}, \ldots, w_{i+l})$ on the *i-th* step of decoding of a *G*-code, generated by the matrixes (34) equals the sum of weights of its coordinates, regarded as code words of a convolution code on the steps $i, i+1, \ldots, i+l$, then the results of decoding the *i-th* step by means of the procedure described above, and of decoding the convolutional (r, s, l) code with the Viterbi decoder would coincide on steps $i, i+1, \ldots, i+l$.

The number of operations and the complexity of each step of a *G*-code decoding is estimated by the value $T_G = O(2^{2k+m})$. At the same time, it follows from the procedure described above that it parallelizes 2^k alternative vectors $x_i \in V_k$ by the outer cycle without any extra input at the transition to the next step of coding. In other words, provided that there are 2^τ parallelizing channels, the complexity of each step of decoding *G*-codes is $T_G = O(2^{2k+m-\tau})$, and in case of maximal parallelization, i.e., when $\tau = k$, we get $T_G = O(2^{k+m})$.

Concluding what has already been said, the following few points need mentioning.

1. Linear convolutional codes are not *G*-codes. And, conversely, *G*-codes are not linear convolutional codes.
2. Any convolutional (r, s, l)-code is presentable as *a* *G*-codes with parameters $n = r(l+1+t), k = s(l+t)$ and $m = s, t \geq 0$.

3. The results of decoding with the Viterbi decoder convolutional codes and of the respective *G*-codes coincide.

Let us note, that, considering everything above, the a priori level of noise-resistance of convolutional codes is not altered by their presentation as respective *G*-codes.

4. MAXIMUM NOISE-RESISTANT *G*-CODES

In this section the algebraic conditions of noise-resistance of *G*-codes are developed, and an algorithm for their construction, based on the outcomes of the first section, is offered. It is proven that the *G*-codes, which are the presentation of convolutional codes, do not fulfill the conditions of maximum noise-resistance.

According to the outcomes of the first section, the decoding of (n,k)-codes is related to the solution of (x,A,δ)-systems. Therefore, the a priori level of noise-resistance of linear codes depends on Hamming weights, the code words, built on the matrix A, and the weights $\chi(\delta)$. If the distance of an (n,k)-code is $d+1$ and $\chi(\delta)\le d/2$, then, as has already been pointed out, a (x,A,δ)-system has an unequivocal solution. If $\chi(\delta)>d/2$, then the a priori probability of an unequivocal solution increases with the decline of the number of code words with minimal Hamming weights - $d+1, d+2$ and so on.

An arbitrary linear (n,k)-code with the distance $d+1$ we will call *maximum noise-resistant,* if:

1. for fixed values of n and k the parameter d is maximal;
2. the number of code words ordered with Hamming weight is $d+1, d+2$ and soon is minimal.

Let us remind ourselves that an optimal (n,k)-code with the distance $d+1$ has a maximal rank k for the given values of n and d. In other words, the optimality of a code means the maximal code rate $R=k/n$ for the given a priori level of noise-resistance, which is defined by *the parameter* d.

So, the best situation will be that when the values of parameters n, k and d simultaneously correspond to both the optimal and the maximum noise-resistant code. As for isomorphic linear codes, the Hamming weights' structure of the families of their code words is the same, building the optimal maximum noise-resistant (n,k)-code means building all the non-isomorphic optimal (n,k)-codes with the maximal possible value of parameter d and then choosing the maximum noise-resistant (by the structure of their code words) Hamming weights.

The decoding procedure for *G*-codes, which has been described in the previous section, is related to the solution of $\left((x',u,x''),\begin{Vmatrix}A\\C\\B\end{Vmatrix},\delta\right)$-systems, where $(x',u,x'')\in V_{2k+m}$ and $\delta\in V_n$. On the *i-th* step of decoding the vectors

$$z_i=(x_{i-1}\overline{\oplus}x_1)A\overline{\oplus}(u_i\overline{\oplus}u_2)C\overline{\oplus}(x_i\overline{\oplus}x_2)B\overline{\oplus}\delta_i \qquad (36)$$

are analyzed for all 2^{2k+m} vectors $(x_1,u_2,x_2)\in V_{2k+m}$ and error vector $\delta_i\in V_n$. The weight function is set by the Hamming weight $\chi(z_i)$ of vectors (36), and the maximum likelihood principle is contained in choosing by minimal respective cumulative Hamming weights.

For $x_2=x_i$ the total system comes down to a $\left((x',u),\begin{Vmatrix}A\\C\end{Vmatrix},\delta\right)$-system, where $(x',u)\in V_{k+m}$. At that, the a priori probability of the situation when the "backward memory" of the G-code decoder would contain exactly the likelihood vector (x_{i-1},u_i), at the address x_i, which depends on the structure of the Hamming weights of the respective code words, would be maximal if the $(n,k+m)$-code, generated by the matrix $\begin{Vmatrix}A\\C\end{Vmatrix}$, is maximum noise-resistant.

Similarly, for $x_1 = x_{i-1}$ we get an $(n, k+m)$-code, generated by the matrix $\begin{Vmatrix} C \\ B \end{Vmatrix}$. If it is maximum noise-resistant, then the a priori probability that at the address $x' \neq x_i$ exactly the likelihood vector (x_{i-1}, u_i) would be written, and at that, the "cumulative weight" for the vector x' would be less than for the likelihood vector x_i, would be minimal.

In the general case when $x_1 \neq x_{i-1}$ and $x_2 \neq x_i$, the maximum noise-resistance of a $(n, 2k+m)$-code, generated by the matrix $\begin{Vmatrix} A \\ C \\ B \end{Vmatrix}$, means the maximal a priori probability that the "current cumulative weights" of vectors $x_2 \in V_k$ would be more than those of the likelihood vector x_i.

Further, the a priority probability of the formation of a loop the length of 1, i.e., the situation when in "backward memory" at the address x_{i+1} the vector (x', u') is written, where $x' \neq x_i$, while at the address x' the likelihood vector (x_{i-1}, u_i) is written, depends on the maximum noise-resistance of the $(2n, k+2m)$-code, generated by the matrix $\begin{Vmatrix} C & \varnothing \\ B & A \\ \varnothing & C \end{Vmatrix}$. For a loop the length of 2 the generating $(2k+3m) \times (3n)$ matrix of the respective code is $\begin{Vmatrix} C & \varnothing & \\ B & A & \\ \varnothing & C & \varnothing \\ & B & A \\ & \varnothing & C \end{Vmatrix}$. For loops the length of $\lambda > 2$ the matrix would be of an analogous echelon form, and its order would be $(\lambda k+(\lambda+1)m)\times((\lambda+1)n)$. Further, we will show that the maximum noise-resistance of the respective codes for loops the length of $\lambda>2$ depends on the maximum noise-resistance of the codes with the generating matrixes, presented above, for the loop lengths $\lambda=1$ and $\lambda=2$.

So, the maximum possible noise-resistance of *G*-codes is achievable when the following codes are maximum noise-resistant:

1. $(n,k+m)$-codes with generating matrixes $\begin{Vmatrix} A \\ C \end{Vmatrix}$ and $\begin{Vmatrix} C \\ B \end{Vmatrix}$.

2. $(n,2k+m)$-codes with generating matrixes $\begin{Vmatrix} A \\ C \\ B \end{Vmatrix}$.

3. $(2n,k+2m)$-code and $(3n,2k+3m)$-code with generating matrixes $\begin{Vmatrix} C & \varnothing \\ B & A \\ \varnothing & C \end{Vmatrix}$ and $\begin{Vmatrix} C & \varnothing & \\ B & A & \\ \varnothing & C & \varnothing \\ & B & A \\ & \varnothing & C \end{Vmatrix}$.

A question arises: how to combine the conditions 1, 2 and 3? In this connection, let us point out that, as opposed to the Viterbi decoder for convolutional codes, in the *G*-code decoder for any vector $x_i \in V_k$ the maximal likelihood vector $(x_{i-1},u_i)\in V_{k+m}$ is chosen from all 2^{k+m} vectors of the set $\left\{xA\overline{\oplus}uC\overline{\oplus}x_iB \,\middle|\, (x,u)\in V_{k+m}\right\}$, and their order in terms of lexicographic numeration does not matter.

Suppose $\pi:V_k\leftrightarrow V_k$ – is some substitution on the set of vectors V_k. Obviously, $\left\{xA\overline{\oplus}uC \,\middle|\, (x,u)\in V_{k+m}\right\}=\left\{\pi(x)A\overline{\oplus}uC \,\middle|\, (x,u)\in V_{k+m}\right\}$. In other words, the matrixes A and B can be substituted by tables with 2^k addresses

$x \in V_k$, in which the sets of vectors $\{\pi_1(x)A \mid x \in V_k\}$ and $\{\pi_2(x)B \mid x \in V_k\}$ are written for any substitutions $\pi_1, \pi_2 : V_k \leftrightarrow V_k$. At that, such substitution has no effect on the results of *G*-codes, given that the coding equation is of the form $y_i = \pi_1(x_{i-1})\overline{A \oplus u_i C \oplus} \pi_2(x_i)B$. Therefore, the substitutions π_1 and π_2 have no effect on the conditions of maximum noise-resistance of *G*-codes 1 and 2, but affect the condition 3.

For vectors $x \in V_k$ let us denote $\chi_A(x) = \min\limits_{u \in V_m} \chi\left((x,u)\left\|\begin{matrix} A \\ C \end{matrix}\right\|\right)$ and $\chi_B(x) = \min\limits_{u \in V_m} \chi\left((u,x)\left\|\begin{matrix} C \\ B \end{matrix}\right\|\right)$. Suppose then, $\chi_{AB}(x_1, x_2) = \min\limits_{u \in V_m} \chi\left((x_1, u, x_2)\left\|\begin{matrix} A \\ C \\ B \end{matrix}\right\|\right)$ for vectors $x_1, x_2 \in V_k$. Considering these denotations, it follows from the form of the generated matrixes for loops the length of $\lambda = 1$ and $\lambda \geq 2$ that the solution to the problem of building the maximum noise-resistant *G*-codes depends on the possibility of finding the substitutions $\pi_1, \pi_2 : V_k \leftrightarrow V_k$, for which the sums of weights $\chi_A(\pi_1(x)) + \chi_B(\pi_2(x)), x \in V_k$, and $\chi_B(\pi_2(x_1)) + \chi_{AB}(\pi_1(x_1), \pi_2(x_2)) + +\ldots + \chi_{AB}(\pi_1(x_{\lambda-1}), \pi_2(x_\lambda)) + \chi_A(\pi_1(x_\lambda)), x_i \in V_k, i = \overline{1,\lambda}$, comply with the definition of a maximum noise-resistant linear (n,k)-code.

For an integer parameter $\rho, 0 \leq \rho \leq n$, let us introduce the denotation $X(\rho) = \{(x_1, x_2) \in V_{2k} \mid \chi_{AB}(x_1, x_2) \leq \rho\}$. The function $|X(\rho)|, 0 \leq \rho \leq n,$ is non-decreasing and the parameter ρ_0 can be found from the condition $|X(\rho_0)| \leq 2^k < |X(\rho_0 + 1)|$.

From what has been said, it therefore follows that the maximum noise-resistance of $(3n, 2k+3m)$-codes from condition 3 for loops of length $\lambda = 2$ mainly depends on the availability of substitution $\pi_1, \pi_2 : V_k \leftrightarrow V_k$, for which the values of the sum $\chi_B(\pi_2(x_1)) + \chi_{AB}(\pi_1(x_1), \pi_2(x_2)) + \chi_A(\pi_1(x_2))$, in particular, for the vectors $(\pi_1(x_1),\pi_2(x_2))\in X(\rho_0)$, for which $1\le \chi_{AB}(\pi_1(x_1),\pi_2(x_2))\le \rho_0$ would be maximal.

Moreover, if the substitutions π_1 and π_2 are such that $(\pi_1(x_1),\pi_2(x_2))\in X(\rho_0)$ and $(\pi_1(x_2),\pi_2(x))\notin X(\rho_0)$ for any vector $x\in V_k$, then the fulfillment of the respective conditions for loops of length $\lambda = 2$ generates their fulfillment for loops of length $\lambda > 2$.

To summarize, the following algorithm for building maximum noise-resistant codes can be suggested.

The algorithm consists of three stages.

1. Suppose $\left\{e_0^{(k)}, x_1, ..., x_{2^k-1}\right\} = V_k$ – is the set of vectors V_k in lexicographical order. Let us define a family of couples of substitutions $\pi_1,\pi_2 : V_k \leftrightarrow V_k$, for which $\chi_A(\pi_1(x_1))\ge \chi_A(\pi_1(x_2))\ge ...\ge \chi_A(\pi_1(x_{2^k-1}))$ and $\chi_B(\pi_2(x_1))\le \chi_B(\pi_2(x_2))\le ...\le \chi_B(\pi_2(x_{2^k-1}))$, as $\Pi_0(\pi_1,\pi_2)$. So, by means of substitutions from the set $\Pi_0(\pi_1,\pi_2)$, the maximal value of the sum $\chi_A(\pi_1(x))+\chi_B(\pi_2(x))$ is achieved for all non-zero vectors $x\in V_k$ with a guarantee of the maximum noise-resistance condition 3 for *G*-codes fulfillment for loops of length $\lambda = 1$. Let us highlight that the values of these sums do not change if the vectors V_k are permuted with the same values of the respective Hamming weights.
2. From the set $\Pi_0(\pi_1,\pi_2)$ of substitutions a subset $\Pi_1(\pi_1,\pi_2)\subseteq \Pi_0(\pi_1,\pi_2)$ of couples of substitutions is chosen, such that their cumulative weights $\chi_B(\pi_2(x_1))+\chi_{AB}(\pi_1(x_1),\pi_2(x_2))+\chi_A(\pi_1(x_2))$ would be maximal for the sets of vectors $(\pi_1(x_1),\pi_2(x_2))\in X(\rho_0)$.

3. From the set of couples of substitutions $\Pi_1(\pi_1,\pi_2)$ a subset $\Pi_2(\pi_1,\pi_2)\subseteq\Pi_1(\pi_1,\pi_2)$ is chosen, for which the condition is fulfilled: if $(\pi_1(x_1),\pi_2(x_2))\in X(\rho_0)$, then $(\pi_1(x_2),\pi_2(x))\notin X(\rho_0)$ for any vector $x\in V_k$.

To summarize, it is worth pointing out that the problem of building of maximum noise-resistant *G*-codes is solvable and practically realizable, if for the given n,k and m it is possible to build the codes that fulfill the 1 and 2 conditions.

So, the solution to the problem of building maximum noise-resistant *G*-codes mainly comes down to the possibility of building optimal maximum noise-resistant $(n,k+m)$-codes with generating matrixes $\begin{Vmatrix} A \\ C \end{Vmatrix}$ and $\begin{Vmatrix} C \\ B \end{Vmatrix}$. Therefore, for the solution of this problem the results, obtained in the first section and presented in Table 2, are crucial.

Let us note that for any given *G*-code it is always possible to build a structure of Hamming weights of the code words that comply with 1, 2 and 3 conditions, and so to estimate its a priori level of noise-resistance, and therefore to compare different *G*-codes by this criterion.

Let us demonstrate that the *G*-codes are the representation of convolutional codes, and do not fulfill the conditions 1 and 2 of maximum noise-resistance of *G*-codes. Indeed, from the form of the matrixes (34) it follows that in the general case linear codes generated by matrixes $\begin{Vmatrix} A_{s(l+t)} \\ C_s \end{Vmatrix}$ and $\begin{Vmatrix} C_s \\ B_{s(l+t)} \end{Vmatrix}$ for any $t\geq 0$ have a distance that does not exceed $r-s$, while the analogous distance for the maximum noise-resistance *G*-codes with the same parameters is limited by $(r-s)(l+1+t)$, $l\geq 1$.

At the same time, the situation with the condition 3 of maximum noise-resistance is somewhat different. In particular, the matrix $\begin{Vmatrix} C_s & \varnothing \\ B_{s(l+t)} & A_{s(l+t)} \\ \varnothing & C_s \end{Vmatrix}$ is, in fact, $l+t+2, t \geq 0$, of the matrix C_s, shifted mutually by r positions. For example, for $l=1$ and $t=0$ it is of the form $\begin{Vmatrix} A_0 & A_1 & A_2 \dots A_l & & \\ & A_0 & A_1 \dots A_{l-1} & A_l & \\ & & A_0 \dots A_{l-2} & A_{l-1} & A_l \end{Vmatrix}$. Obviously, the distance of a linear code, generated by such a matrix, depends on the properties of the matrix C_s. Moreover, these properties would be preserved not only for the loops of length $\lambda=1$, but also for $\lambda \geq 2$. Let us highlight that this particular property makes traditional convolutional codes quite effective. At the same time, this confirms the fact that the a priori level of noise-resistance of *G*-codes with matrixes (34) coincides with the respective level of the initial convolutional code and is not altered for any $t \geq 0$. Having said that, the a priori level of noise-resistance for maximum noise-resistant *G*-codes with parameters $n=r(l+1+t), k=s(l+t)$ and $m=s$ is not only initially higher than that of convolutional codes, but also increases with the increase of parameter t.

In summary, it is appropriate to point out the following issues.

1. The classical problem of the solution of systems of linear equations with skewed right-hand side, or of decoding linear codes, allows a matroid interpretation both in terms of its statement and solution methods realization. This fact enables us to use the theoretical outcomes obtained for matroids in general and for semi-matroids in particular, for the solution of this type of problem.
2. Based on the outcomes of the first section, an algorithm for building the maximum noise-resistant *G*-codes is offered, which allows us to optimize the choice of values of parameters n, k and m of *G*-code generating matrixes for particular systems of communication.
3. The substitution of traditional linear convolutional (r, s, l)-codes by the maximum noise-resistant G-codes with parameters $n=r(l+1+t), k=s(l+t)$ and $m=s, t \geq 0$, and the set of matrixes (A_t, C_t, B_t) guarantees the increase of the information transfer efficiency. Along with that, the systems of communication can use both the G-codes with generating matrixes

$(A_{s(l+t)}, C_s, B_{s(l+t)})$, which correspond to the initial convolutional code, and the maximum noise-resistant G-codes. For this it is sufficient to replace the matrixes A_t, C_t, B_t with the matrixes $A_{s(l+t)}, C_s, B_{s(l+t)}$, and vice versa in the same coding procedure.

***

Thus, in the fourth part it has been shown that the apparatus of pseudo-matroids and semi-matroids, introduced in the previous parts, can be applied to the solution of a range of practical problems. In particular, the problem on building generating matrixes for all non-isomorphic optimal linear codes for quite a wide spectrum of values of parameter $|S|$. Based on this, the maximum noise-resistant *G*-codes are built, the use of which in communication systems, along with classical convolutional codes, allows for the increase of noise-resistance of such systems without any alteration of their characteristics.

CONCLUSION

In the category of matroids and their mappings every matroid $M \in \mathfrak{M}(S)$ can be regarded as a result of a canonical mapping $B(S) \to M$. A canonical mapping, like any other strong mapping, is factorized into elementary mappings. Elementary mappings, in turn, are unequivocally defined by some modular cuts of the family of flats or by linear cuts of the family of co-points of the respective factors of the free matroid $B(S)$. So, hypothetically, any matroid from the set $\mathfrak{M}(S)$ is defined as some sequence of modular cuts of the family of flats of a free matroid.

To build a matroid means to describe its family of cycles, bases and co-points. The problem is that the systems of axioms, which unequivocally define the properties of a matroid, are not sufficient to build a matroid. In other words, by means of these axioms, it cannot be checked whether a given group of subsets of the set S is the family of cycles, bases or co-points of some matroid from the set $\mathfrak{M}(S)$. This means that the actual system of axioms does not contain any procedure for building such groups of sets, other than enumeration. Let us highlight that this particular issue makes the problem of the enumeration of all matroids, including non-isomorphic in the set $\mathfrak{M}(S)$ for the sets S of relatively low cardinality, rather difficult. At the same time, the fact that it is actually possible to describe a matroid by the sequence of its modular cuts and modular filters of a free matroid $B(S)$ transfers the problem of building all of the matroids from the set $\mathfrak{M}(S)$ into the area of analysis of the properties of all the respective modular cuts and filters. At that, obviously, it is necessary to abstract oneself from the axiomatization of modular cuts as some subset of flats of a matroid that is built based on mutual axiomatic systems.

The notion of pseudo-matroids, generated by the elementary mappings of matroids, introduced in this study, is the actual instrument which enables us to study the properties of modular filters from a point of view that differs from the classical one.

The family of cycles of a matroid, generated by an elementary mapping $M \xrightarrow{\Phi;\mathfrak{Y}} H$ of matroids $M, H \in \mathfrak{M}(S)$, coincides with the family of minimal by inclusion elements of the modular filter Φ_M, and the family of bases – with the set of co-points from the family $\mathfrak{K}_M$, that does not belong to the modular cut $\mathfrak{Y}_M$. Thus, the family of all elementary factors of any matroid M is unequivocally defined by the respective family of pseudo-matroids. At the same time, it is necessary to mention that it is not generally possible to set this family axiomatically as some family of algebraic objects generated by the matroid M. For example, for Higg's factorization, the families of cycles and bases of a pseudo-matroid coincide respectively with the sets $\mathfrak{R}_H$ и $\mathfrak{B}_H$ of the matroid H and cannot be expressed through analogous structures of the matroid M. For top reduction, on the contrary, these families coincide with the sets $\mathfrak{B}_M$ and $\mathfrak{K}_M$ of the matroid M and do not define the form of the matroid H. The investigation of the properties of pseudo-matroids, generated by the elementary mappings of matroids, brought about a range of general theoretical outcomes. For instance, under a canonical mapping of $B(S) \to M$, the family of Higgs lifts of a matroid M is expressed explicitly through the family of cycles $\mathfrak{R}_M$. The criterion of erectability for matroids from the set $\mathfrak{M}(S)$ has been obtained.

Also, in this study a new notion for the theory of matroids – that of G-mapping – the elementary mapping of a certain type between binary matroids from the set $\mathfrak{N}(S)$ has been introduced. A special case of pseudo-matroids – semi-matroids, generated by the G-mapping of binary matroids from the set $\mathfrak{N}(S)$, is evaluated. A characteristic difference between the case of semi-matroids and the general case is the possibility of axiomatic definition of the abovementioned family of pseudo-matroids, generated by any binary matroid. It is proven that for any binary matroid $M \in \mathfrak{N}(S)$ there always exists a factorization of a canonical mapping $B(S) \to M$ into G-mappings, and the category of binary matroids is closed under morphisms – G-mappings and rank-preserving weak mappings. Thus, all binary matroids can be built, as in the general case, by means of some sequence of semi-matroids. However, as opposed to pseudo-matroids, the respective procedure not only is not an enumeration, but also allows us to explicitly construct the families of cycles of the respective G-factors. Moreover, the application of semi-matroid apparatus allowed the suggestion of an algorithm for building the families of cycles of all non-isomorphic binary matroids from the set $\mathfrak{N}(S)$ with the computational complexity of $O\left(|S|^3 3^{|S|}\right)$ arithmetical operations, which means a solution of the problem of enumeration of all non-isomorphic binary matroids for the set S of a rather high cardinality.

Finally, it is proven that for any binary matroid it is possible to construct a family of axiomatically defined semi-matroids, which would coincide with the set of all semi-matroids, generated by its G-factors. This, in turn, means that the family of all such axiomatically set semi-matroids, generated by binary matroids from the set $\mathfrak{N}(S)$ can be considered as an independent algebraic structure, functionally connected with the category of binary matroids and their mappings.

It is worth mentioning that the apparatus of pseudo-matroids and semi-matroids, introduced in this work, can serve as a basis for further study of the properties of the category of matroids and their mappings.

In the final part of this work general theoretical outcomes are applied for the solution of a range of practical problems. In the first section, the transformed algorithm for building all non-isomorphic binary matroids is also used for building non-isomorphic binary matroids with given restrictions on their families of cycles. This fact, considering the respective decrease of computational complexity of the algorithm, allows us to actually build the generating matrixes of all non-isomorphic linear codes for a rather wide range of values of parameter $|S|$.

In the second section, a range of issues related to the problem of finding the solution for systems of linear equations with skewed right-hand side or decoding of linear codes. It is demonstrated that this problem has a matroid equivalent and, therefore, allows the matroid interpretation of related problems. In the third and fourth sections a method of noise-resistant G-coding is considered along with an algorithm for building maximum noise-resistant G-codes, based on the outcomes of the first section. It is proven that classical convolutional codes are presentable as *G*-codes, which have a lower level of noise-resistance than the maximum possible. By means of these *G*-codes, systems of communication can be created that not only guarantee a higher noise-resistance, compared to the maximum possible of convolutional codes, but also allow the selection of those *G*-codes that optimally comply with the characteristics of particular systems of communication.

REFERENCES

[1] Birkhoff, G. Lattice theory//*Amer. Math*, 1967.

[2] Crapo, HH. Erecting geometrics//*Anuals N.Y. Acad. Sci.*, 1970 P. 89-92.

[3] Crapo, HH; Rota, GG. On the foundations of Combinatorial Theory. Combinatorial Geometrics. Cambridge, Mass.: M.I.T. Press, 1970.

[4] Dowling, TA; Kelly, DG. Elementary strong maps and transversal geometries//*Discrete Math.*, 1974. Vol. 7. P. 209-214.

[5] Higgs, DA. Maps of geometries//*J. London Math. Soc.*, 1966. Vol. 41 P. 612-618.

[6] Higgs, DA. Strong maps of geometries//*J. Comb. Theory.*, 1968. Vol. 5. P. 185-191.

[7] Kelly, PJ; Kennedy, DG. The Higgs factorization of a geometric strong map//*Discrete Math.*, 1978. Vol. 22. N. 2.P. 139-146.

[8] Kennedy, DG. Factorization and majors of geometric strong maps. Lecture Note Series. Univ. North Carolina, 1973.

[9] Lucas, D. Properties of rank preserving weak maps//*Bull. Amer. Math. Soc.*, 1974. Vol. 80. N. 1. P. 127-131.

[10] Nguyen, HQ. Functors of the category of combinatorial geometries and strong map//*Discrete Math.*, 1977. Vol. 20. N. 2. P. 143-158.

[11] Nguyen, HQ. Weak cuts of combinatorial geometries//*Trans. Amer. Math.*, 1979. vol. 250.

[12] Oxley, JG. Matroid Theory. N.Y.: Oxford Univ. Press, 1992. P. 532.

[13] Tutte, WT. Introduction to the theory of matroids. N.Y.: Amer. Elsevire, 1970.

[14] Welsh, DJA. Matroid Theory. L.: Acad. Press, 1976. P. 433.

[15] Whitney, H. On the abstract properties of linear dependence//*Amer. J. Math.*, 1935. Vol. 57. P. 509-533.

[16] Gizunov, SA; Lyamin, VN. Cuts of matroids//Uchen. Zapiski. *Electronic journal of Kursk state university*., 2010. N. 2(14). P. 21. http://scientific-notes.ru.

[17] Gizunov, SA; Lyamin, VN. Pseudo-matroids, generated by mapping of matroids // Uchen. Zapiski. *Electronic journal of Kursk state university*., 2011. N. 2(18). P. 26. http://scientific-notes.ru.

[18] Gizunov, SA; Lyamin, VN. Semi-matroids, generated by *G*-mappings of binary matroids//Uchen. Zapiski. *Electronic journal of Kursk state university*., 2011. N. 2(18). P. 26. http://scientific-notes.ru.

[19] Gizunov, SA; Grechkin, AO; Lyamin, VN. G-factorization of canonical mappings//Uchen. Zapiski. *Electronic journal of Kursk state university*., 2011. N. 2(18). P. 15. http://scientific-notes.ru.

[20] Gizunov, SA. Non-isomorphic binary matroids//Uchen. Zapiski. *Electronic journal of Kursk state university*., 2011. N. 2(18). P. 18. http://scientific-notes.ru.

[21] Gizunov, SA. Optimal linear codes and the critical problem for matroids. //Uchen. Zapiski. *Electronic journal of Kursk state university*., 2011. N. 2(18). P. 19. http://scientific-notes.ru.

[22] Revyakin, AM. The description of combinatory geometry's factors. Combinatory analysis.//*M.: MSU*, 1976. Issue 4. P. 69-72.

[23] Revyakin, AM. Matroids: cryptomorphic systems of axioms and strict firms.//Vestnik MGADA. *Series "Philosophical, humanity and social sciences"*, 2010. N. 5. P. 96-106.

AUTHOR CONTACT INFORMATION

Dr. Sergey A. Gizunov
Doctor of Sciences
Professor
Scientific Research Institute "KVANT"
Moscow, Russia
Email: sa_gizunov@rambler.ru

Dr. N. V. Lyamin, PhD
Scientific Research Institute "KVANT"
Moscow, Russia
Email: nv_lyamin@rambler.ru

INDEX

A

B

C

D

E

F

G

L

M

N

O

P

R

S